Zika Virus and Diseases

Zika Virus and Diseases

From Molecular Biology to Epidemiology

Suzane Ramos da Silva
Fan Cheng
Shou-Jiang Gao

Department of Molecular Microbiology and Immunology,
Keck School of Medicine, University of Southern California,
Los Angeles, California, USA

The right of Suzane Ramos da Silva, Fan Cheng and Shou-Jiang Gao to be identified
as the author(s) of this work has been asserted in accordance with law.

Registered Offices
John Wiley & Sons, Inc., 111 River Street, Hoboken, NJ 07030, USA

Editorial Office
111 River Street, Hoboken, NJ 07030, USA

For details of our global editorial offices, customer services, and more information
about Wiley products visit us at www.wiley.com.

Wiley also publishes its books in a variety of electronic formats and by print-on-demand.
Some content that appears in standard print versions of this book may not be available
in other formats.

Limit of Liability/Disclaimer of Warranty
The publisher and the authors make no representations or warranties with respect to
the accuracy or completeness of the contents of this work and specifically disclaim
all warranties; including without limitation any implied warranties of fitness for a
particular purpose. This work is sold with the understanding that the publisher is not
engaged in rendering professional services. The advice and strategies contained herein
may not be suitable for every situation. In view of on-going research, equipment
modifications, changes in governmental regulations, and the constant flow of
information relating to the use of experimental reagents, equipment, and devices, the
reader is urged to review and evaluate the information provided in the package insert
or instructions for each chemical, piece of equipment, reagent, or device for, among
other things, any changes in the instructions or indication of usage and for added
warnings and precautions. The fact that an organization or website is referred to in
this work as a citation and/or potential source of further information does not mean
that the author or the publisher endorses the information the organization or website
may provide or recommendations it may make. Further, readers should be aware that
websites listed in this work may have changed or disappeared between when this
work was written and when it is read. No warranty may be created or extended by any
promotional statements for this work. Neither the publisher nor the author shall be
liable for any damages arising here from.

Library of Congress Cataloging-in-Publication data applied for

Hardback: 9781119408642

Cover Design: Wiley
Cover Images: ©Igor Normann / Shutterstock; ©jipatafoto89 / Shutterstock;
©Kateryna Kon / Shutterstock; ©Mike Goldwater / Alamy Stock Photo

Set in 10/12pt Warnock by SPi Global, Pondicherry, India

Printed in Singapore by C.O.S. Printers Pte Ltd

10 9 8 7 6 5 4 3 2 1

Contents

Preface

Zika virus (ZIKV), a mosquito-borne flavivirus, is an emerging infectious agent associated with numerous neurological diseases. Discovered in 1947, the virus was silent for almost 60 years until the recent outbreaks in 2003 and 2013 in the Pacific Islands, and in 2015 in South and Central America. While the virus detected in Africa at the time of discovery was only associated with mild fever, rash, and pain, recent ZIKV outbreaks were associated with neurological disorders such as Guillain-Barré Syndrome in adults and microcephaly in newborns. The dynamic changes in ZIKV-associated pathology over the years has prompted extensive studies aimed at understanding the differences among the virus lineages (African vs. Asian/American) isolated from different regions with the goal of developing specific therapeutic drugs.

This book will describe the ZIKV story since its discovery in 1947 up to the most updated studies in 2017. We will cover more than 70 years of ZIKV history with details in the discovery, outbreaks, transmission, associated diseases, animal models that have been developed, ZIKV and cell/host interactions, the differences among ZIKV strains, and drugs that have been tested against ZIKV. This book should provide valuable information for both the general public and scientists.

List of Abbreviations

Abbreviations	Full name
2′-CMA	2′-C-methyladenosine
2′-CMC	2′-C-methylcytidine
2′-CMG	2′-C-methylguanosine
2′-CMU	2′-C-methyluridine
2′-O-Me	2′-O ribose methylation
3-MA	3-methyladenine
7-deaza-2′-CMA	7-deaza-2′-C-methyladenosine
7DMA	7-deaza-2′-C-methyladenosine
aa	amino acid
AEN	apoptosis enhancing nuclease
AIM	absent in melanoma
ALKBH	Alkylation repair homologs
AMPK	5′ adenosine monophosphate-activated protein kinase
Atg	autophagy-related protein
ATP	adenosine triphosphate
BPTI	bovine pancreatic trypsin inhibitor
BUNV	Bunyamvera virus
C	capsid protein
CCL	Chemokine (C-C motif) ligand
CCN	cyclin
CD	cluster of differentiation
CDK	cyclin-dependent kinase
CDKi	CDK inhibitor
CDKN	cyclin dependent kinase inhibitor
CFS	cerebrospinal fluid

CHIKV	chikungunya virus
CM	conditioned medium
C_{max}	maximum plasma concentrations
CPAP	centrosomal P4.1-associated protein
CPE	cytopathic effect
CRISPR	Clustered regularly interspaced short palindromic repeats
CXCL	Chemokine (C-X-C motif) ligand
CXCR	C-X-C motif chemokine receptor
D.P.C.	days post-coitus
D.P.I.	days post-infection
DC	dendritic cell
DC-SIGN	Dendritic cell-specific intercellular adhesion molecule-3-grabbing non-integrin
DENV	dengue virus
DISC	death-inducing signaling complex
DNA	deoxyribonucleic acid
dsRNA	double-stranded RNA
E	envelope protein
EBV	Epstein-Barr virus
ELISA	enzyme-linked immunosorbent assay
ELISA	enzyme-linked immunosorbent assay
ER	endoplasmic reticulum
FDA	Food and Drug Administration
FFU	focus forming units
FLE	fusion loop epitope
FLUAV	influenza A virus
FLUBV	influenza B virus
fNPC	fetal neural progenitor cell
fNSC	fetal NSC
FTO	fat mass and obesity-associated protein
GEF	guanine nucleotide exchange factor
GM	genetically modified
GO	gene ontology
GTP	guanosine-5′-triphosphate
H2AX	H2A histone family, member X
HAEC	human amnion epithelial cell
HC	Hofbauer cell
HCV	hepatitis C virus

hESC	human embryonic stem cell
HI	hemagglutination-inhibition
hiPSC	human inducible pluripotent stem cell
HIV	human immunodeficiency virus
hNPC	human neural progenitor cell
hnRNP	heterogeneous nuclear ribonucleoprotein
hpi	hour(s) post infection
HSV	herpes simplex virus
I.C.	intracerebrally
I.P.	intraperitoneally
IFIT	IFN-induced proteins with tetratricopeptide repeats
IFITM	IFN-inducible transmembrane protein
IFN	interferon
IKK	IκB kinase
IL	interleukin
IP-10	Interferon gamma-induced protein 10
IPS	interferon-promoter stimulator
iPSC	induced pluripotent stem cells
IRF	IFN regulatory factor
ISG	IFN-stimulated gene
ISRE	IFN-stimulated responsive element
IUGR	intrauterine growth restriction
JEV	Japanese encephalitis virus
KUNV	Kunjin virus
LC3	microtubule-associated protein 1A/1B-light chain 3
LDH	lactate dehydrogenace
LGP2	laboratory of genetics and physiology 2
LGTV	Langat virus
M	membrane protein
m^6A	N^6 methylation of adenosine
MAVS	Mitochondrial antiviral-signaling protein
MAYV	Mayaro virus
MBFV	mosquito-borne flaviviruses virus
MCM	Mauritian cynomolgus macaque
MCP1	monocyte chemoattractant protein-1
MDA 5	melanoma-differentiation-associated gene 5
MDE	mean day of euthanasia
MEF	mouse embryonic fibroblast
METTL	methyltransferase-like

MHC	major histocompatibility complex
MLD	mucin-like domain
moDC	monocyte-derived DC
MOI	multiplicity of infection
MORF	MOZ-related factor
MOZ	monocytic leukemic zinc-finger protein
MPA	mycophenolic acid
MR	monkey rhesus
MRI	magnetic resonance imaging
mRNA	messanger RNA
MTase	methyltransferase
mTORC	mammalian target of rapamycin complex
MyD88	myeloid differentiation primary response gene 88
NCX-NES cells	neocortical NES cells
NES cells	neuroepithelial stem cells
NF-κB	nuclear factor-κB
NKV	no known vector
NMR	nuclear magnetic resonance
NPC	neural progenitor cell
NPC	neural progenitor cells
NS	nonstructural protein
NSC	neural stem cell
NTPase	nucleoside triphosphatase
OAS	2′-5′-oligoadenylate synthetase
ORF	open reading frame
p.i.	post-infection
PAMP	pathogen-associated molecular pattern
PARP	poly-ADP ribose polymerase
PAS	pre-autophagosomal structure
PCD	programmed cell death
PCNT	Pericentrin
PE	phosphatidylethanolamine
PFU	plaque-forming units
PG	phosphatidylglycerol
PHT	primary human trophoblast
PI3K	phosphatidylinositol-3-kinase
PKR	protein kinase R
PKA	protein kinase A
PKI	PKA inhibitor
pNGB	p-nitrophenyl-p-guanidino benzoate

PQS	potential quadruplex sequence
prM	the precursor of membrane protein
PRNT	plaque reduction neutralization tests
$PRNT_{50}$	50% plaque reduction neutralizing titer
PRR	pattern recognition receptor
PTEN	phosphatase and tensin homolog
PTK	protein tyrosine kinase
qPCR	quantitative polymerase chain reaction
RANTES	regulated on activation, normal T cell expressed and secreted
RdRp	RNA-dependent RNA polymerase
RF	replicative form
RGC	radial glia cell
RGP	radial glial progenitor
RI	replicative intermediate
RIG-I	retinoid-inducible gene I
RLR	RIG-I-like receptors
RNA	ribonucleic acid
ROS	reactive oxygen species
RPE cells	retinal pigment epithelial cells
RSAD	radical S-adenosyl methionine domain containing
RSP	recombinant subviral particle
RTPase	RNA triphosphatase
RT-PCR	reverse-transcriptase polymerase chain reaction
RT-qPCR	quantitative reverse transcription PCR
S.C.	subcutaneously
SAM	S-adenosyl-methionine
SARS	severe acute respiratory syndrome
SD	standard deviation
sfRNA	subgenomic flaviviral RNA
SINV	Sindbis virus
SPOV	Spondweni virus
ssRNA	single stranded RNA
STAT	signal transducer and activator of transcription
STING	stimulator of interferon gene
TAM	Tyro3-Axl-Mer
TANK	TRAF family member-associated NF-κB activator
TBFV	tick-borne flavivirus
TBK	TANK-binding kinase
TEM	transmission electron microscopy

TIM	T-cell immunoglobin and mucin domain
TIR	Toll/interleukin-1 receptor
TLR	Toll-like receptor
t_{max}	duration to achieve C_{max}
TMD	transmembrane domain
TMEV	Theiler's mouse encephalomyelitis virus
TNF	tumor necrosis factor
TRAF	tumor necrosis factor receptor-associated factor
TRAIL	TNF-related apoptosis inducing ligand
TRIF	TIR-domain-containing adapter-inducing interferon-β
TSC	tuberous sclerosis
ULK1	Unc-51-like kinase
UTR	untranslated region
VEEV	Venezuelan equine encephalitis virus
VP	vesicle packet
WDR	WD40 repeat
WNV	West Nile virus
WS	Webster Swiss
xrRNA	Xrn1-resistent RNA
YF	yellow fever
YFV	yellow fever virus
YTH	YT521-B homology
YTHDF	YTH N6-methyladenosine RNA binding protein
ZFYVE	zinc finger FYVE-type
ZIKV	Zika virus
ZIKV^{BR}	ZIKV strain isolated in Brazil
γH2AX	phosphorylated histone H2AX

1

The History of ZIKV Discovery

1.1 ZIKV Isolation from Monkeys and Mosquitos

Zika virus (ZIKV) was first isolated in April 1947 in a forest named "Ziika" near Lake Victoria in Uganda (1, 2). It is interesting to note that the term *Ziika* means "overgrowth" in Luganda (the Bantu language of the Baganda people, commonly used in Uganda). The virus was isolated by researchers from The National Institute for Medical Research in London, United Kingdom (G. W. A. Dick), and The Rockefeller Foundation in New York, United States (S. F. Kitchen and A. J. Haddow), as part of collaborative studies with the Yellow Fever Research Institute in Entebbe, Uganda (Figure 1.1) (1, 2).

To monitor emerging infections, the investigators commenced studying the sentinel rhesus monkeys in Bwamba, Uganda, in 1946 (Figure 1.2) (1). Zika Forest was chosen because it was well-known that monkeys in that area had a high immunity to yellow fever virus (YFV) (3–6). Most of the forest was parallel to the Entebbe-Kampala Road, and the monkeys were kept in cages in the canopy of the trees (1, 7–9).

At that time, six monkey rhesus (MR) were monitored daily for any variation in their body temperature. One of the monkeys, named MR766, presented an increase in temperature on April 18; hence, a blood sample was collected on April 20. MR766 was monitored for more 30 days but no other symptom was detected. The blood sample collected from MR766 was injected subcutaneously (S.C.) into another monkey named MR771, and into Swiss albino mice, intracerebrally (I.C.) and intraperitoneally (I.P.), for further studies. No sign of

Zika Virus and Diseases: From Molecular Biology to Epidemiology, First Edition.
Suzane Ramos da Silva, Fan Cheng and Shou-Jiang Gao.
© 2018 John Wiley & Sons, Inc. Published 2018 by John Wiley & Sons, Inc.

Figure 1.1 Alexander J. Haddow in the Zika Forest. The base of the platform used to capture mosquitoes and keep the monkeys can be observed. Obtained from the University of Glasgow (AJ Haddow and University of Glasgow Archives & Special Collections, Papers of AJ Haddow, GB248 DC 068/80/63).

infection was observed in either MR771 or the mice injected by I.P. for up to 30 days after inoculation. However, the mice injected by I.C. became sick 10 days post-infection (d.p.i.), and the first ZIKV isolation was obtained from these animals. Since this virus was neutralized by serum taken from monkeys MR766 (on May 20) and MR771 (at 35 d.p.i.) but not by sera from these same monkeys before their exposure to ZIKV, the researchers proved that the virus isolated from the mice was originated from monkey MR766. For this reason, the first ZIKV strain isolated was named MR766. A neutralizing antibody is the antibody that can protect the cells from an infection by neutralizing its biological effect (in this case, infection). In this study, it was used in an assay to determine if the virus detected in one animal was the same as the one isolated from the previous animal (1).

In addition to analyzing and collecting samples from the monkeys, the researches were collecting mosquitos for the YF studies (Figure 1.3). *Aedes africanus* were among the captured ones in 1948. This mosquito was suspected to be involved in the YFV cycle at that time. From January 5 to January 20, nine lots of mosquitos were

Figure 1.2 Details of the steel tower used as a platform to collect mosquitos, and to keep the caged monkeys in the Zika Forest. The platforms can reach the canopy of the trees. Obtained from the University of Glasgow (AJ Haddow and University of Glasgow Archives & Special Collections, Papers of AJ Haddow, GB248 DC 068/80/62).

acquired, and their samples were processed and injected into mice by I.C. with both unfiltered supernatants and Seitz E.K. filtrates. The second isolation of ZIKV (strain E/1), which was also the first from mosquitos, was from lot E/1/48, captured on January 11–12, with 86 mosquitos (1). All six mice inoculated with the unfiltered sample were inactive at 7 d.p.i. For animals that received the filtrated sample, one died at 6 d.p.i. while other was sick at 14 d.p.i.

Figure 1.3 Details on the stairs used to access and recover the mosquitoes caught. A boy can be observed in the picture, since they were used to help the researchers to collect the samples in the high height. Obtained from the University of Glasgow (AJ Haddow and University of Glasgow Archives & Special Collections, Papers of AJ Haddow, GB248 DC 068/80/49).

Those inoculates were also injected S.C. into MR758, which showed no sign of sickness. Based on the results observed with the sick mice, blood samples from MR758 were collected on the 8th, 9th, and 10th d.p.i., which were I.C. injected into six mice. From the first injection, one mouse died at 10 d.p.i. and another two became sick at 19 and 20 d.p.i. From the second group of injection, one died at 13 d.p.i., one was sick at 12 d.p.i., and another one developed paralysis, which was identified as Theiler's encephalomyelitis (10, 11). Mice injected with samples from the third collection had no symptom. Neutralization tests with serum from MR758 proved that these animals had developed neutralization antibodies to ZIKV strains E/1 and MR766. It was concluded that ZIKV was identical to neither YFV, Dengue virus (DENV), nor Theiler's mouse encephalomyelitis virus (TMEV) (1).

Dick (1952) observed that the virus isolated from MR766 and mosquitos was well adapted after 90 passages in the mouse brain. Data from studies analyzing adaption and pathogenesis became more reproducible. Among the three ZIKV strains tested (MR766, MR758, and E/1), the virus from MR758 caused more cases of mice that presented with paralysis in early passage than the virus from MR766. With all the strains evaluated, the first sign of infection was roughness of the coat. Infection by I.P. injection in mice older than 2 weeks of age was not as efficient in those of 7 days old. Using a late-passage virus, no significant difference in the infection was observed between unweaned and 5- to 6-week-old mice (2).

The virus tropism was examined by analyzing infection in different organs, including brain, kidney, lung, liver, and spleen. The results of the mice inoculated by I.C. indicated that the brain was the main target of ZIKV. While other animals including cotton-rat and guinea pigs could also be infected with ZIKV, no symptom was observed. On the other hand, rabbits could produce antibodies by 21 d.p.i. Other species of monkeys—including rhesus (6 animals), grivet (13 animals), and redtails (2 animals)—were also infected and analyzed. Circulating antibodies were detected in Grivet 733 and Redtail 1044 after ZIKV infection. Interestingly, Grivet 1019 was naturally infected by ZIKV, but this monkey was captured in Sese Island, which was not in the Zika region. In 1950, among the monkey rhesus used for the YFV research, animal MR801 was naturally positive for ZIKV but the only symptom was minor pyrexia. MR801 was kept in the same platform (number 3) where the strain E/1 was isolated from the captured mosquitoes. Platform number 3 was 0.2 miles from platform number 5 where MR766 was infected (2). Antibodies against ZIKV were not detected in small mammals that were trapped in the forest, indicating that the infection was restricted to monkeys, mosquitos, and human beings at that time (12, 13).

Other ZIKV strains were isolated in 1958 from two different catches of *Aedes africanus*, consisting of 206 (strain Lunyo V) and 127 (strain Lunyo VI) mosquitoes. The Lunyo V strain caused viral encephalitis, skeletal myositis, and myocarditis in adults and infant mice. The virus was passed through the brain and the heart into infant mice via I.C. or I.P. injections. Some of the mice injected with Lunyo VI were paralyzed. The strains were injected into monkeys MR1059 and MR1063, respectively, and no symptoms of infection were observed (14). ZIKV was further isolated from *Aedes luteocephalus* in Nigeria (15).

1.2   ZIKV Infection in Humans

The timing of the first ZIKV infection in humans is controversial (16). A paper published by MacNamara in 1954 described its isolation and exploited the possible association between ZIKV infection and jaundice (17). Another study, by Bearcroft in 1956, was on a volunteer that self-injected with the virus, who precisely described the symptoms following the infection (18). The only problem is that the virus isolated in the first study and used in the second one was not ZIKV but a Spondweni virus (SPOV) (16). MacNamara's study evaluated an epidemic of jaundice in Nigeria (Afikpo Division, Eastern Nigeria). From a study of three patients, the virus was isolated from a 10-year-old female patient who was not jaundiced but had symptoms of fever and headache, and her serological response to ZIKV was low (17).

Bearcroft's study was done to verify whether there was any association between ZIKV and the development of jaundice. A 34-year-old European male volunteer was exposed to the virus isolated by MacNamara (1956). Eighty-two hours post-infection (h.p.i.), the only symptom was a headache, followed by malaise and pyrexia in the 2nd and 3rd d.p.i. In the 5th d.p.i., there was a peak in the corporal temperature, accompanied by nausea and vertigo, which was diagnosed as histamine reaction. After 7 days, the volunteer had no sign of infection or jaundice. Mice infected with virus collected from the volunteer, in different periods, developed encephalitis after receiving sera collected at 4 and 6 d.p.i., which were around the peak of pyrexia. Meanwhile, the volunteer was exposed to *Aedes aegypti*, but the mosquitos were not able to transmit the infection to infant white mice (18).

The first clue that both studies were using SPOV was revealed in a study by Simpson (1964), which was also the first one to describe a natural infection of ZIKV in humans (19). In this paper he mentioned that previous isolations of ZIKV were made in Nigeria (West Africa), and Dr. Delphine Clarke had found out that those viruses were closely related to SPOV, which was named CHUKU strain. Another study in

1968 also pointed out that SPOV virus was isolated in Nigeria, and was wrongly identified as ZIKV (20). Simpson was actually the person who contracted the infection while working together with his team in Uganda. A detailed description of his symptoms following the natural infection was recorded. At the 1 d.p.i., he presented a headache, and by 2 d.p.i., he developed a red rash diffused throughout his face, neck, chest, and arms, without itching, and slight pain in the back and thighs. The rash covered all the limbs, including palms and soles. The fever started at 2 d.p.i., followed by malaise. At 3 d.p.i., the patient had no fever and did not feel sick, and at 5 d.p.i., there was no more rash (19). Actually, this was the first study that documented the presence of a rash on humans infected by ZIKV, one of the most common symptoms of ZIKV infection in today's patients (21).

The first isolation of ZIKV in Nigeria was reported in 1975 by Moore (1975) in a study describing the isolation of 15 arboviruses between 1964 and 1970 (22). Isolation of ZIKV in Oyo State, Nigeria, was described in 1979. The virus was isolated from two patients, a 2½-year-old boy with a mild fever in 1971, and a 10-year-old male in 1975, who presented with fever, headache, and pain in the body. This study suggested that ZIKV might be widespread, even if it had been isolated at a low rate. One important point mentioned in this study was that ZIKV infection numbers might be underestimated because of the mild symptoms or misdiagnosis with other arthropod-borne viral infections (23).

1.3   ZIKV Infection Spread to Other Hosts and Regions

Different serological studies were performed around the 1950s and 1960s and showed that the ZIKV infection had reached other areas in Africa and Asia (24, 25). Serological analysis, based on hemagglutination-inhibition (HI) tests of other animals, were described in 1977 with samples from 2,428 small mammals and 1,202 birds captured over a five-year period in Kano Plain, Kenya, close to Lake Victoria. The results revealed the prevalence of ZIKV antibodies as follows:

- In small mammals:
    - 4.0% (58/1,446) in *Arvicanthus niloticus*
    - 34.0% in (85/250) *Arvicanthis niloticus*
    - 3.1% (2/63) in *Crocidura occidentalis*

- In reptiles:
  - 40.0% (4/10) in *Boaedon fuliginosus*
  - 12.5% (1/8) in *Varanus niloticus*
- in birds:
  - 4.0% (2/49) in *Threskiornis aethiopicus*
  - 2.7% (1/37) in *Bubulcus ibis* and 50.0% (1/2) in *Philomachus pugnax*
- In other mammals:
  - 0.1% (1/655) in goats
  - 0.7% (2/283) in sheep
  - 0.6% (8/1361) in cattle living close to irrigated areas
  - 0.7% (7/963) in cattle from nonirrigated places (26)

Serological studies with human serum collected for the YFV research indicated that humans from some areas were exposed to ZIKV. There was no detection of ZIKV antibodies in the Zika and Kampala regions, while Bwamba had detection rates of ZIKV antibodies at 28.5% (2/7) in adults and 15.4% (2/13) in children, which were higher than the 9.5% (2/21) detection rate of West Nile antibodies in adults in this region (2). Dick (1952) was careful in his study and suggested that just because there was no evidence of an acute disease in humans caused by ZIKV infection, this did not indicate that ZIKV was not important or might not cause any problem in humans (2).

The detection of antibodies against ZIKV in South-East Asia was published in 1963, revealing that 75.0% (from 100 samples) of the population living in the Federation of Malayan (currently known as Peninsula of Malaya) was positive, while the presence of neutralization antibodies in the north region such as North Vietnam and Thailand (Bangkok and Chiang Mai) was rare (27). In 1965, ZIKV was detected in different regions of the Angola trough with 31.0% (40/129) and 1.5% (2/129) rates in children in the northwestern region by HI and neutralization tests, respectively, and with 57.7% (71/123) and 21.1% (26/123) rates in adults, respectively, by the same methods. In the southwestern region, 32.8% (20/61) and 21.3% (13/61) of the adults were positive by HI and neutralization tests, respectively, and for the eastern region, 3.5% (2/56) and 1.8% (1/56) of the adults were positive, also using HI and neutralization tests, respectively (28). Results from Kano Plain, Kenya, showed that ZIKV was endemic in 1973, but it was considered at a low level. By analyzing sera from children (ages 4–15+ years old) from schools distributed close to Lake Victoria,

ZIKV was detected by HI test in 7.2% (40/559) of the children grouped as 12 years old. Since this was considered a low rate, the other ages were not evaluated (29).

In 1974, a serological study to detect different arboviruses analyzed 1,649 human sera from Portugal and identified four (0.25%) individuals that reacted against ZIKV by the HI test, indicating the silent spread of the virus across the continents (30). In 1979, a serological study analyzed 235 samples from Hong Kong and detected 4.6% (11/235) ZIKV-positive individuals, which also cross-reacted with other flaviviruses. Among those who had gender and age information, 12.9% (4/31) females and 8.3% (1/12) males between 21 and 40 years old were positive, while 7.1% (1/14) males older than 41 years old were positive (31).

Interesting results were found at Kainji Lake Basin, Nigeria, in 1980, when ZIKV was detected by HI test, with cases concentrated in young adults and adults. Infection rate was correlated with increased age. Specifically, 9.3% (7/75) of 5- to 9-year-olds, 22.2% (8/36) of 10- to 14-year-olds, 46.1% (6/13) of 15- to 19-year-olds, 71.8% (61/85) of 20- to 39-year-olds, and 77.3% (68/88) of adults 40 years old and older (32) were positive. The continuous ZIKV detection by the HI test throughout Uganda villages in 1984 indicated that the incidence of ZIKV was not common in the region, with infection rates at 3.7% (1/27) in Tokora, 15.4% (2/13) in Nadip, 3.5% (2/58) in Namalu, and 20.0% (3/15) in other regions (33).

1.4 Cross-Paths between ZIKV and Other Flaviviruses

Analysis of sera collected from two different towns, Ilaro and Ilobi, in southwest Nigeria in 1951 and 1955 showed high detection rates of ZIKV antibodies in these populations. The distribution of ZIKV infection by age was as follows: 10.0% (3/29) among 5- to 9-year-olds, 22.0% (7/32) among 10- to 14-year-olds, 52.0% (13/25), among 15- to 19-year-olds, 76.0% (19/25) among 20- to 29-year-olds, 52.0% (16/31) among 30- to 39-year-olds, and 93.0% (28/30) for those adults 40 years old and older in Ilobi. In Ilaro, only children samples (4 to 16 years old) were collected, which showed a 44.0% positive rate for ZIKV antibodies. Besides ZIKV, high infection rates were also detected for DENV, YFV,

Uganda S virus (UGSV), and Bwamba fever virus (BWAV). There was association of antibodies against ZIKV, DENV, YFV, and UGSV, suggesting an overlapping protection. However, infection by one virus did not decrease the chance of being infected by another flavivirus, albeit it might reduce the pathogenesis. The most common combinations of infections were ZIKV or UGSV with YFV. Hence, a pre-infection with either ZIKV or UGSV might produce neutralization antibodies to YFV. ZIKV had a strong association with UGSV, YFV, and DENV. The DENV infection rate reached close to 100% in this region. It was suggested that ZIKV and UGSV might suppress YFV in the Forest Belt compared to other regions with high incidences of YFV and lower infection rates of ZIKV and UGSV (34).

The cyclic periodicity between ZIKV and chikungunya virus (CHIKV) was suggested by McCrae (1971) because there was no evidence that both viruses were maintained in the Entebbe region, but there were epizootic outbreaks with intervals of 5 to 8 and up to 10 years. The intervals were similar but not the same, with ZIKV following CHIKV outbreak after 1 to 2 years, which might be the result of the dynamic interactions of the viruses within the forest (35). In 1982, a study was published to address the possible interaction and interference in transmission between ZIKV and YFV. Mosquito's catches resulted in the isolation of 15 ZIKV strains from *Aedes africanus* and *Aedes apicoargenteus*. Dozens of monkeys were shot (after unsuccessful attempts of collecting blood) and captured to provide evidence of immunity change among the monkeys in the forest. Twenty-two Redtail monkeys were captured in Kisubi Forest while 68 monkeys (including Redtail, Mona, Colobus, and Mangabeys) were caught in Bwamba. After serological analysis, CHIKV was detected at high rates among the monkeys followed by ZIKV and YFV with similar rates, indicating that ZIKV infection did not prevent the circulation of YFV (36).

References

1 Dick GW, Kitchen SF, Haddow AJ. 1952. Zika virus. I. Isolations and serological specificity. Trans R Soc Trop Med Hyg **46**:509–520.

2 Dick GW. 1952. Zika virus. II. Pathogenicity and physical properties. Trans R Soc Trop Med Hyg **46**:521–534.

3 Haddow AJ, Dick GW, Lumsden WH, Smithburn KC. 1951. Monkeys in relation to the epidemiology of yellow fever in Uganda. Trans R Soc Trop Med Hyg **45**:189–224.

4 Findlay GM, Maccallum FO. 1937. Yellow fever immune bodies in the blood of African primates. Transactions of the Royal Society of Tropical Medicine and Hygiene **31**:103–106.

5 Maccallum FO, Findlay GM. 1937. Yellow fever immune bodies and animal sera. Transactions of the Royal Society of Tropical Medicine and Hygiene **31**:199–206.

6 Haddow AJ, Smithburn KC, et al. 1947. Monkeys in relation to yellow fever in Bwamba County, Uganda. Trans R Soc Trop Med Hyg **40**:677–700.

7 Smithburn KC, Kerr JA, Gatne PB. 1954. Neutralizing antibodies against certain viruses in the sera of residents of India. J Immunol **72**:248–257.

8 Smithburn KC. 1954. Neutralizing antibodies against arthropod-borne viruses in the sera of long-time residents of Malaya and Borneo. Am J Hyg **59**:157–163.

9 Hammon WM, Schrack WD, Jr., Sather GE. 1958. Serological survey for a arthropod-borne virus infections in the Philippines. Am J Trop Med Hyg **7**:323–328.

10 Theiler M. 1934. Spontaneous Encephalomyelitis of Mice–a New Virus Disease. Science **80**:122.

11 Theiler M. 1937. Spontaneous Encephalomyelitis of Mice, a New Virus Disease. J Exp Med **65**:705–719.

12 Haddow AJ, Williams MC, Woodall JP, Simpson DI, Goma LK. 1964. Twelve Isolations of Zika Virus from Aedes (Stegomyia) Africanus (Theobald) Taken in and above a Uganda Forest. Bull World Health Organ **31**:57–69.

13 Dick GW. 1953. Epidemiological notes on some viruses isolated in Uganda; Yellow fever, Rift Valley fever, Bwamba fever, West Nile, Mengo, Semliki forest, Bunyamwera, Ntaya, Uganda S and Zika viruses. Trans R Soc Trop Med Hyg **47**:13–48.

14 Weinbren MP, Williams MC. 1958. Zika virus: further isolations in the Zika area, and some studies on the strains isolated. Trans R Soc Trop Med Hyg **52**:263–268.

15 Lee VH, Moore DL. 1972. Vectors of the 1969 yellow fever epidemic on the Jos Plateau, Nigeria. Bull World Health Organ **46**:669–673.

16 Wikan N, Smith DR. 2017. First published report of Zika virus infection in people: Simpson, not MacNamara. Lancet Infect Dis **17**:15–17.

17 Macnamara FN. 1954. Zika virus: a report on three cases of human infection during an epidemic of jaundice in Nigeria. Trans R Soc Trop Med Hyg **48**:139–145.

18 Bearcroft WG. 1956. Zika virus infection experimentally induced in a human volunteer. Trans R Soc Trop Med Hyg **50**:442–448.

19 Simpson DI. 1964. Zika Virus Infection in Man. Trans R Soc Trop Med Hyg **58**:335–338.

20 Boorman JP, Draper CC. 1968. Isolations of arboviruses in the Lagos area of Nigeria, and a survey of antibodies to them in man and animals. Trans R Soc Trop Med Hyg **62**:269–277.

21 Musso D, Gubler DJ. 2016. Zika Virus. Clin Microbiol Rev **29**:487–524.

22 Moore DL, Causey OR, Carey DE, Reddy S, Cooke AR, Akinkugbe FM, David-West TS, Kemp GE. 1975. Arthropod-borne viral infections of man in Nigeria, 1964–1970. Ann Trop Med Parasitol **69**:49–64.

23 Fagbami AH. 1979. Zika virus infections in Nigeria: virological and seroepidemiological investigations in Oyo State. J Hyg (Lond) **83**:213–219.

24 Smithburn KC, Taylor RM, Rizk F, Kader A. 1954. Immunity to certain arthropod-borne viruses among indigenous residents of Egypt. Am J Trop Med Hyg **3**:9–18.

25 Smithburn KC. 1952. Neutralizing antibodies against certain recently isolated viruses in the sera of human beings residing in East Africa. J Immunol **69**:223–234.

26 Johnson BK, Chanas AC, Shockley P, Squires EJ, Gardner P, Wallace C, Simpson DI, Bowen ET, Platt GS, Way H, Parsons J, Grainger WE. 1977. Arbovirus isolations from, and serological studies on, wild and domestic vertebrates from Kano Plain, Kenya. Trans R Soc Trop Med Hyg **71**:512–517.

27 Pond WL. 1963. Arthropod-Borne Virus Antibodies in Sera from Residents of South-East Asia. Trans R Soc Trop Med Hyg **57**:364–371.

28 Kokernot RH, Casaca VM, Weinbren MP, McIntosh BM. 1965. Survey for antibodies against arthropod-borne viruses in the sera of indigenous residents of Angola. Trans R Soc Trop Med Hyg **59**:563–570.

29 Bowen ET, Simpson DI, Platt GS, Way H, Bright WF, Day J, Achapa S, Roberts JM. 1973. Large scale irrigation and arbovirus epidemiology, Kano Plain, Kenya. II. Preliminary serological survey. Trans R Soc Trop Med Hyg **67**:702–709.

30 Filipe AR. 1974. Serological survey for antibodies to arboviruses in the human population of Portugal. Trans R Soc Trop Med Hyg **68**:311–314.

31 Johnson BK, Chanas AC, Gardner P, Simpson DI, Shortridge KF. 1979. Arbovirus antibodies in the human population of Hong Kong. Trans R Soc Trop Med Hyg **73**:594–596.

32 Adekolu-John EO, Fagbami AH. 1983. Arthropod-borne virus antibodies in sera of residents of Kainji Lake Basin, Nigeria 1980. Trans R Soc Trop Med Hyg **77**:149–151.

33 Rodhain F, Gonzalez JP, Mercier E, Helynck B, Larouze B, Hannoun C. 1989. Arbovirus infections and viral haemorrhagic fevers in Uganda: a serological survey in Karamoja district, 1984. Trans R Soc Trop Med Hyg **83**:851–854.

34 Macnamara FN, Horn DW, Porterfield JS. 1959. Yellow fever and other arthropod-borne viruses; a consideration of two serological surveys made in South Western Nigeria. Trans R Soc Trop Med Hyg **53**:202–212.

35 McCrae AW, Henderson BE, Kirya BG, Sempala SD. 1971. Chikungunya virus in the Entebbe area of Uganda: isolations and epidemiology. Trans R Soc Trop Med Hyg **65**:152–168.

36 McCrae AW, Kirya BG. 1982. Yellow fever and Zika virus epizootics and enzootics in Uganda. Trans R Soc Trop Med Hyg **76**:552–562.

2

ZIKV: From Silent to Epidemic

Up to the 1980s, studies investigating flaviviruses had only paid special attention to yellow fever virus (YFV), mostly focusing on African and Asian countries. Because Zika virus (ZIKV) infection had mostly been associated with fever, headache, and cutaneous rash up to the time of the recent outbreaks, studies focusing on ZIKV had never become the focus of the mainstream scientific community since its discovery in 1947 (1–3).

2.1   Outbreak in Yap Island (2007)

In 2009, Duff et al. (4) reported the first outbreak of ZIKV outside African and Asian countries (5–10). In 2007, physicians in Yap Island (Federated States of Micronesia) reported that many residents were presenting serious symptoms, which included cutaneous rash, conjunctivitis, slight fever, arthralgia, and arthritis (4). Even after some of the sera from these patients tested positive for IgM antibodies against dengue virus (DENV) using an available commercial kit, the clinical findings revealed that the new outbreak was not the same as those typically associated with DENV infection. Evaluation for other flaviviruses was done by reverse-transcriptase polymerase chain reaction (RT-PCR) assay, and ZIKV was identified in some (7/71; 14%) of the sera. Analysis of medical health center data between April and July 2007 allowed the identification of 185 patients that presented symptoms similar to those described above. From these patients, 26.5% were confirmed to have the ZIKV

Zika Virus and Diseases: From Molecular Biology to Epidemiology, First Edition.
Suzane Ramos da Silva, Fan Cheng and Shou-Jiang Gao.
© 2018 John Wiley & Sons, Inc. Published 2018 by John Wiley & Sons, Inc.

disease; 31.9% were classified as possible ZIKV disease; 38.9% were suspected to have ZIKV diseases but no serum was collected in the acute phase of infection (up to 10 days after onset of the symptoms); and 2.7% were not infected by ZIKV. To be classified as a confirmed case of ZIKV infection, the serum from the patient had to be positive for ZIKV RNA by RT-PCR, or had to present three additional parameters: IgM antibodies against ZIKV by enzyme-linked immunosorbent assay (ELISA) (11); ZIKV titer ≥ 20 in plaque reduction neutralization tests ((cutoff value of 90%) ($PRNT_{90}$) (12)); and ratio of ≥ 4 for ZIKV $PRNT_{90}$/DENV $PRNT_{90}$ (4). If the ratio between ZIKV and DENV $PRNT_{90}$ was lower than 4, and no RNA was detected by RT-PCR in the serum (or the serum was not available for evaluation), then the case could only be classified as probably infected by ZIKV (4, 11).

A survey of 200 randomly selected households was conducted to estimate the infection rate among the 1,276 households living in Yap Island. The members of the families were evaluated if they were at least 3 years old. From the 557 individuals evaluated, 414 (74.0%) had IgM antibodies against ZIKV and of those 414, 156 had symptoms like those caused by ZIKV infection. Of the 143 individuals who did not have IgM antibodies against ZIKV, 27 reported feeling sick with symptoms like ZIKV infection. Using these findings to extrapolate the infection rate among the Yap Island residents, it was concluded that 73.0% of the population had been infected by ZIKV in 2007, which was the first outbreak reported in the island. The water-holding containers for the population were also evaluated, and 87.0% (148/170) of the mosquitos or larva pupae were ZIKV positive. Twelve different species of mosquitos were identified. The predominant one was *Aedes hensilli* (4). Although it was not possible to isolate ZIKV from these samples, sequence alignment inferred that this strain was not the same as the African ones previously described, and was denominated as Asian strains (11).

2.2 Outbreak in French Polynesia (2013)

A case involving a 3-year-old boy who presented with 4 days of fever, sore throat, cough, and 3 days of headache was reported in Cambodia in 2010. The boy was positive for ZIKV by PCR, sequencing, and serology, and negative for DENV and Chikungunya virus (CHIKV). This was an isolated case (13). Since then, there has been no outbreak in Cambodia, with a detection of only low prevalence in the population (14).

Following the outbreak in the Yap Island, French Polynesia, which is located in the South Pacific and composed of 119 islands, a routine detection for ZIKV infection by RT-PCR has been implemented (15). At week 41 of 2013, three patients from the same household presented symptoms that were consistent with ZIKV infection. Patient 1 was a 53-year-old woman, her 52-year-old husband was patient 2, while patient 3 was their 42-year-old son-in-law. They all had slight fever, asthenia, wrist and finger arthralgia, headache, and a rash. The male patients also presented conjunctivitis while the female had swollen ankles and aphthous ulcers. The tests were negative for DENV, CHIKV, and West Nile virus (WNV) by RT-PCR. Results for ZIKV were equivocal for patients 1 and 2. At week 43, a 57-year-old patient with similar symptoms was tested negative for DENV but positive for ZIKV by RT-PCR. Between weeks 43 and 44, the Department of Health registered an increase of patients reporting similar ZIKV infection symptoms to their primary doctors. Ten samples were tested and 4 (40.0%) were positive for ZIKV. The virus isolated from these samples presented high homology with those isolated from Cambodia in 2010 and Yap Island in 2007. At week 51, a total of 5,895 patients had suspected ZIKV infection. Of 584 samples analyzed, 294 (50.3%) were positive for ZIKV, indicating that an outbreak of ZIKV had reached French Polynesia (15).

Since ZIKV usually presented mild symptoms of infection, a survey was conducted to analyze the proportion of symptomatic and asymptomatic individuals who had ZIKV infection in French Polynesia. In this survey, samples from 196 patients from the five archipelagoes of French Polynesia, 700 samples (September to October of 2015) from Tahiti and Moorea, and 476 samples from children (May to June of 2014) from Tahiti were collected. The detection of IgM antibodies to ZIKV based on ELISA demonstrated that ZIKV infection in symptomatic and asymptomatic individuals were 49.0% and 43.0% for the first group; 22.0% and 53.0% for the second group; and 6.0% and 29.0% for the children (16). Another study with samples from October 2013 to March 2014 showed a low rate (210/747; 28.1%) of ZIKV RNA detection in sera of symptomatic (in acute phase) patients by RT-PCR, indicating that other samples, such as urines, should be used to confirm cases of ZIKV infection before dismissing any patients (17).

After the outbreak, some cases of imported ZIKV infection were detected in other countries. Japan had two cases imported from French Polynesia, a male in his mid-20s, which presented fever, headache, arthralgia, and rash. The second was his partner, a female in her early 30s, which presented retro-orbital pain, slight fever, rash, and itching. His blood serum was positive for ZIKV and hers was negative but her urine was positive. Sequence analysis showed that the virus was highly similar to the ones from Cambodia in 2010 and Yap Island in 2007 but differed from the ones isolated in Africa (18).

After 2014, a bloom of imported infections was described, which was mostly concentrated among Pacific Islands, Asia and Europe, including cases from Cook Island to Australia (19), from Indonesia to Australia (20, 21), from Thailand to Japan (22), from Thailand to Canada (23), from Senegal to the United States (24), from Thailand to Germany (25), from Tahiti to Norway (26), and from Malaysia to Germany (27). Subsequent outbreaks of ZIKV infection were described in Easter Island (28), New Caledonia (29), American Samoa (30), and other Pacific Islands (31), sometimes simultaneously with outbreaks of other flaviviruses (32). Other countries, including Singapore (33–37), Thailand (38, 39), and Philippines (40) also reported outbreaks with local transmission of ZIKV.

2.3   How Did ZIKV Reach Brazil?

How ZIKV reached South America is still debatable. It was first assumed that it was due to the World Cup soccer championship held in Brazil in July of 2014 (41), but the Pacific Islands that reported the ZIKV outbreak were not competing. Interestingly, in August of 2014, there was a Va'a World Sprint Championship canoe race in Brazil, where teams from French Polynesia, New Caledonia, Cook Islands, and Easter Island were participating (42). Another possible opportunity of ZIKV introduction into Brazil was during the Fifa Confederations Cup in 2013. A recent sequencing study has concluded that ZIKV was detected in Brazil in early 2014.

2.4   Outbreak in Brazil (2015)

The first autochthonous transmission described in Brazil was in the beginning of 2015 in the city of Natal located in the Northeast state, Rio Grande do Norte. Sera from 21 patients with a "dengue-like syndrome" were collected in the acute phase. The patients presented symptoms that were similar to those of DENV and CHIKV infections, including pain, myalgia, headache, and slightly fever. All cases were negative for DENV and CHIKV while 38.0% of the cases were positive for ZIKV by RT-PCR. For the positive cases, 87.5% were from female patients, and all patients presented rash and pain. Sequence alignment of the virus demonstrated that the strain in Brazil was 99.0% similar to the one isolated in French Polynesia in 2013, belonging to the Asian branch (41).

Bahia, another state in the Northeast region, also reported an outbreak. In March 2015, 24 patients with "dengue-like syndrome" were tested for DENV, ZIKV, CHIKV, WNV, and Mayaro virus (MAYV). All samples were negative for DENV, WNV, and MAYV while three samples were positive for CHIKV, and seven other samples were positive for ZIKV by RT-PCR. Most of the patients presented rash, slight fever, myalgia, and headache (44). In Pernambuco, based on the clinical evaluation of the received cases from January to April 2015, 1,046 cases were analyzed for symptoms like ZIKV infection, such as exanthema, pruritus, fever, and arthralgia. Of those, 895 (86.0%) cases were classified as probable cases of ZIKV infection while 151 (14.0%) as DENV cases (45). Detection of ZIKV, DENV, and CHIKV circulating in the same period of time in Pernambuco was also reported (46).

Other regions in Brazil also reported ZIKV infection. In Rio de Janeiro located in the southeast region, 364 patients were suspected to be infected by ZIKV based on clinical analysis from January to July 2015. Samples were evaluated for ZIKV, and 45.4% (119/262) were positive by RT-PCR. The study indicated that the peak of infection was between May and June 2015. The usual symptoms were headache, arthralgia, myalgia, nonpurulent conjunctivitis, lower back pain, and pruritus. Analysis of the virus sequence confirmed that the ZIKV strain belonged to the Asian strain reported in other states (47). Besides the rash, which was detected in almost all patients with ZIKV infection in different studies, most of the patients were women in their third and fourth decade of life (48, 49).

Based on ZIKV detection in different states, the Brazilian Ministry of Health recognized ZIKV as a circulating virus in Brazil, and the Salvador Epidemiologic Surveillance Office registered 14,835 cases described as acute exanthematous illness between February 2015 and June 2015. Some of these cases were serologically screened for different viruses, since DENV and CHIKV had caused outbreaks in the same region in previous years, and 58 were tested positive for DENV, CHIKV, and ZIKV by RT-PCR. The samples were 5.2% positive for ZIKV, 5.2% for CHIKV, 1.7% for DENV-3, and 1.7% for DENV-4. Although there was no indication that ZIKV was the only virus causing the illness, the symptoms matched the ones described for ZIKV infection, especially because of the slight fever and the fact that all patients presented with rash or exanthema. Other symptoms were pruritus in most of the patients, arthralgia, headache, and myalgia (50).

2.5 ZIKV Spread through South, Central, and North Americas

Cases of ZIKV infection in other countries after traveling to Brazil started to emerge in the second decade of the twenty-first century. The first case was a man from Italy who had visited Salvador in the end of March 2015 (51). Since then, ZIKV cases have been detected in other countries in South and Central America such as Colombia (52), Mexico (53), Haiti (54, 55), Puerto Rico (56–58), Guatemala (59), Ecuador (60), Honduras (61), and Martinique (62), to name a few. In Colombia, ZIKV was associated with the death of four febrile patients with age varying from 2 to 72 years old in both genders. From September 2015 to March 2016, Colombia had confirmed, by laboratory tests, 2,361 cases of ZIKV infection (64). Suriname also reported a few fatal cases related to ZIKV infection (63).

A drastic expansion of ZIKV occurred between 2007 and 2017, with local transmission reported in over 84 countries and territories worldwide by March 2017. In February 2016, the World Health Organization (WHO) declared a Public Health Emergency of International concern due to ZIKV infection (65), which ended in November 2016. WHO listed that the last country to report ZIKV detection in 2016 was Papua New Guinea, and the first one to report in 2017 was India. Recent sequencing studies have reported that ZIKV had circulated for months undetected in the Americas before it was detected (66), and the United States had multiple introductions of ZIKV (67).

Figure 2.1, extracted from Bueno et al. (2016), shows the distribution of ZIKV infection worldwide through 2016 according to the timeline. It shows countries that have been affected by ZIKV outbreaks, together with ZIKV detection in humans (colorful areas). The animal sketches indicate which animals have been infected by ZIKV in these areas (68).

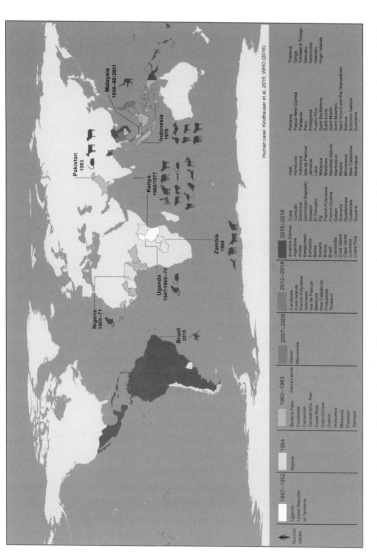

Figure 2.1 Map showing the distribution of ZIKV infection in humans (colorful areas) and animals (sketches) following the chronological outbreaks in different regions. *Source:* Figure and legend extracted from Bueno et al. (2016) (68). (*See insert for color representation of the figure.*)

References

1 Dick GW. 1952. Zika virus. II. Pathogenicity and physical properties. Trans R Soc Trop Med Hyg **46**:521–534.

2 Dick GW, Kitchen SF, Haddow AJ. 1952. Zika virus. I. Isolations and serological specificity. Trans R Soc Trop Med Hyg **46**:509–520.

3 Musso D, Gubler DJ. 2016. Zika Virus. Clin Microbiol Rev **29**:487–524.

4 Duffy MR, Chen TH, Hancock WT, Powers AM, Kool JL, Lanciotti RS, Pretrick M, Marfel M, Holzbauer S, Dubray C, Guillaumot L, Griggs A, Bel M, Lambert AJ, Laven J, Kosoy O, Panella A, Biggerstaff BJ, Fischer M, Hayes EB. 2009. Zika virus outbreak on Yap Island, Federated States of Micronesia. N Engl J Med **360**:2536–2543.

5 Simpson DI. 1964. Zika Virus Infection in Man. Trans R Soc Trop Med Hyg **58**:335–338.

6 Moore DL, Causey OR, Carey DE, Reddy S, Cooke AR, Akinkugbe FM, David-West TS, Kemp GE. 1975. Arthropod-borne viral infections of man in Nigeria, 1964–1970. Ann Trop Med Parasitol **69**:49–64.

7 Fagbami AH. 1979. Zika virus infections in Nigeria: virological and seroepidemiological investigations in Oyo State. J Hyg (Lond) **83**:213–219.

8 Olson JG, Ksiazek TG, Suhandiman, Triwibowo. 1981. Zika virus, a cause of fever in Central Java, Indonesia. Trans R Soc Trop Med Hyg **75**:389–393.

9 Johnson BK, Chanas AC, Gardner P, Simpson DI, Shortridge KF. 1979. Arbovirus antibodies in the human population of Hong Kong. Trans R Soc Trop Med Hyg **73**:594–596.

10 Adekolu-John EO, Fagbami AH. 1983. Arthropod-borne virus antibodies in sera of residents of Kainji Lake Basin, Nigeria 1980. Trans R Soc Trop Med Hyg **77**:149–151.

11 Lanciotti RS, Kosoy OL, Laven JJ, Velez JO, Lambert AJ, Johnson AJ, Stanfield SM, Duffy MR. 2008. Genetic and serologic properties of Zika virus associated with an epidemic, Yap State, Micronesia, 2007. Emerg Infect Dis **14**:1232–1239.

12 Calisher CH, Karabatsos N, Dalrymple JM, Shope RE, Porterfield JS, Westaway EG, Brandt WE. 1989. Antigenic relationships between flaviviruses as determined by cross-neutralization tests with polyclonal antisera. J Gen Virol **70** (Pt 1):37–43.

13 Heang V, Yasuda CY, Sovann L, Haddow AD, Travassos da Rosa AP, Tesh RB, Kasper MR. 2012. Zika virus infection, Cambodia, 2010. Emerg Infect Dis **18**:349–351.

14 Duong V, Ong S, Leang R, Huy R, Ly S, Mounier U, Ou T, In S, Peng B, Ken S, Buchy P, Tarantola A, Horwood PF, Dussart P. 2017. Low Circulation of Zika Virus, Cambodia, 2007–2016. Emerg Infect Dis **23**:296–299.

15 Cao-Lormeau VM, Roche C, Teissier A, Robin E, Berry AL, Mallet HP, Sall AA, Musso D. 2014. Zika virus, French polynesia, South pacific, 2013. Emerg Infect Dis **20**:1085–1086.

16 Aubry M, Teissier A, Huart M, Merceron S, Vanhomwegen J, Roche C, Vial AL, Teururai S, Sicard S, Paulous S, Despres P, Manuguerra JC, Mallet HP, Musso D, Deparis X, Cao-Lormeau VM. 2017. Zika Virus Seroprevalence, French Polynesia, 2014–2015. Emerg Infect Dis **23**.

17 Musso D, Rouault E, Teissier A, Lanteri MC, Zisou K, Broult J, Grange E, Nhan TX, Aubry M. 2016. Molecular detection of Zika virus in blood and RNA load determination during the French Polynesian outbreak. J Med Virol doi:10.1002/jmv.24735.

18 Kutsuna S, Kato Y, Takasaki T, Moi M, Kotaki A, Uemura H, Matono T, Fujiya Y, Mawatari M, Takeshita N, Hayakawa K, Kanagawa S, Ohmagari N. 2014. Two cases of Zika fever imported from French Polynesia to Japan, December 2013 to January 2014 [corrected]. Euro Surveill **19**.

19 Pyke AT, Daly MT, Cameron JN, Moore PR, Taylor CT, Hewitson GR, Humphreys JL, Gair R. 2014. Imported zika virus infection from the cook islands into australia, 2014. PLoS Curr **6**.

20 Leung GH, Baird RW, Druce J, Anstey NM. 2015. Zika Virus Infection in Australia Following a Monkey Bite in Indonesia. Southeast Asian J Trop Med Public Health **46**:460–464.

21 Kwong JC, Druce JD, Leder K. 2013. Zika virus infection acquired during brief travel to Indonesia. Am J Trop Med Hyg **89**:516–517.

22 Shinohara K, Kutsuna S, Takasaki T, Moi ML, Ikeda M, Kotaki A, Yamamoto K, Fujiya Y, Mawatari M, Takeshita N, Hayakawa K, Kanagawa S, Kato Y, Ohmagari N. 2016. Zika fever imported from Thailand to Japan, and diagnosed by PCR in the urines. J Travel Med **23**.

23 Fonseca K, Meatherall B, Zarra D, Drebot M, MacDonald J, Pabbaraju K, Wong S, Webster P, Lindsay R, Tellier R. 2014. First case of Zika virus infection in a returning Canadian traveler. Am J Trop Med Hyg **91**:1035–1038.

24 Foy BD, Kobylinski KC, Chilson Foy JL, Blitvich BJ, Travassos da Rosa A, Haddow AD, Lanciotti RS, Tesh RB. 2011. Probable non-vector-borne transmission of Zika virus, Colorado, USA. Emerg Infect Dis **17**:880–882.

25 Tappe D, Rissland J, Gabriel M, Emmerich P, Gunther S, Held G, Smola S, Schmidt-Chanasit J. 2014. First case of laboratory-confirmed Zika virus infection imported into Europe, November 2013. Euro Surveill **19**.

26 Waehre T, Maagard A, Tappe D, Cadar D, Schmidt-Chanasit J. 2014. Zika virus infection after travel to Tahiti, December 2013. Emerg Infect Dis **20**:1412–1414.

27 Tappe D, Nachtigall S, Kapaun A, Schnitzler P, Gunther S, Schmidt-Chanasit J. 2015. Acute Zika virus infection after travel to Malaysian Borneo, September 2014. Emerg Infect Dis **21**:911–913.

28 Tognarelli J, Ulloa S, Villagra E, Lagos J, Aguayo C, Fasce R, Parra B, Mora J, Becerra N, Lagos N, Vera L, Olivares B, Vilches M, Fernandez J. 2016. A report on the outbreak of Zika virus on Easter Island, South Pacific, 2014. Arch Virol **161**:665–668.

29 Dupont-Rouzeyrol M, O'Connor O, Calvez E, Daures M, John M, Grangeon JP, Gourinat AC. 2015. Co-infection with Zika and dengue viruses in 2 patients, New Caledonia, 2014. Emerg Infect Dis **21**:381–382.

30 Healy JM, Burgess MC, Chen TH, Hancock WT, Toews KE, Anesi MS, Tulafono RT, Jr., Mataia MA, Sili B, Solaita J, Whelen AC, Sciulli R, Gose RB, Uluiviti V, Hennessey M, Utu F, Nua MT, Fischer M. 2016. Notes from the Field: Outbreak of Zika Virus Disease—American Samoa, 2016. MMWR Morb Mortal Wkly Rep **65**:1146–1147.

31 Craig AT, Butler MT, Pastore R, Paterson BJ, Durrheim DN. 2017. Acute flaccid paralysis incidence and Zika virus surveillance, Pacific Islands. Bull World Health Organ **95**:69–75.

32 Roth A, Mercier A, Lepers C, Hoy D, Duituturaga S, Benyon E, Guillaumot L, Souares Y. 2014. Concurrent outbreaks of dengue, chikungunya and Zika virus infections—an unprecedented epidemic wave of mosquito-borne viruses in the Pacific 2012–2014. Euro Surveill **19**.

33 Dyer O. 2016. Outbreak of Zika in Singapore sparks warnings in neighbouring countries. BMJ **354**:i4740.

34 Fisher D, Cutter J. 2016. The inevitable colonisation of Singapore by Zika virus. BMC Med **14**:188.

35 Leo YS, Chow A. 2016. Zika virus has arrived in Singapore. Lancet Infect Dis **16**:1317–1319.

36 Sadarangani SP, Hsu LY. 2016. The 2016 Outbreak of Zika in Singapore. Ann Acad Med Singapore **45**:381–382.

37 Xu BY, Low SG, Tan RT, Vasanwala FF. 2016. A case series of atypical presentation of Zika Virus infection in Singapore. BMC Infect Dis **16**:681.

38 Buathong R, Hermann L, Thaisomboonsuk B, Rutvisuttinunt W, Klungthong C, Chinnawirotpisan P, Manasatienkij W, Nisalak A, Fernandez S, Yoon IK, Akrasewi P, Plipat T. 2015. Detection of Zika Virus Infection in Thailand, 2012-2014. Am J Trop Med Hyg **93**:380–383.

39 Wikan N, Suputtamongkol Y, Yoksan S, Smith DR, Auewarakul P. 2016. Immunological evidence of Zika virus transmission in Thailand. Asian Pac J Trop Med **9**:141–144.

40 Alera MT, Hermann L, Tac-An IA, Klungthong C, Rutvisuttinunt W, Manasatienkij W, Villa D, Thaisomboonsuk B, Velasco JM, Chinnawirotpisan P, Lago CB, Roque VG, Jr., Macareo LR, Srikiatkhachorn A, Fernandez S, Yoon IK. 2015. Zika virus infection, Philippines, 2012. Emerg Infect Dis **21**:722–724.

41 Zanluca C, Melo VC, Mosimann AL, Santos GI, Santos CN, Luz K. 2015. First report of autochthonous transmission of Zika virus in Brazil. Mem Inst Oswaldo Cruz **110**:569–572.

42 Musso D. 2015. Zika Virus Transmission from French Polynesia to Brazil. Emerg Infect Dis **21**:1887.

43 Faria NR, Quick J, Claro IM, Thézé J, de Jesus JG, Giovanetti M, Kraemer MUG, Hill SC, Black A, da Costa AC, Franco LC, Silva SP, Wu CH, Raghwani J, Cauchemez S, du Plessis L, Verotti MP, de Oliveira WK, Carmo EH, Coelho GE, Santelli ACFS, Vinhal LC, Henriques CM, Simpson JT, Loose M, Andersen KG, Grubaugh ND, Somasekar S, Chiu CY, Muñoz-Medina JE, Gonzalez-Bonilla CR, Arias CF, Lewis-Ximenez LL, Baylis SA, Chieppe AO, Aguiar SF, Fernandes CA, Lemos PS, Nascimento BLS, Monteiro HAO, Siqueira IC, de Queiroz MG, de Souza TR, Bezerra JF, Lemos MR, Pereira GF, Loudal D, Moura LC, Dhalia R, França RF, et al. 2017. Establishment and cryptic transmission of Zika virus in Brazil and the Americas. Nature **546**:406–410.

44 Campos GS, Bandeira AC, Sardi SI. 2015. Zika Virus Outbreak, Bahia, Brazil. Emerg Infect Dis **21**:1885–1886.

45 Brito CA, Brito CC, Oliveira AC, Rocha M, Atanasio C, Asfora C, Matos JD, Lima AS, Albuquerque MF. 2016. Zika in Pernambuco: rewriting the first outbreak. Rev Soc Bras Med Trop **49**:553–558.

46 Pessoa R, Patriota JV, Lourdes de Souza M, Felix AC, Mamede N, Sanabani SS. 2016. Investigation Into an Outbreak of Dengue-like Illness in Pernambuco, Brazil, Revealed a Cocirculation of Zika, Chikungunya, and Dengue Virus Type 1. Medicine (Baltimore) **95**:e3201.

47 Brasil P, Calvet GA, Siqueira AM, Wakimoto M, de Sequeira PC, Nobre A, Quintana Mde S, Mendonca MC, Lupi O, de Souza RV, Romero C, Zogbi H, Bressan Cda S, Alves SS, Lourenco-de-Oliveira R, Nogueira RM, Carvalho MS, de Filippis AM, Jaenisch T. 2016. Zika Virus Outbreak in Rio de Janeiro, Brazil: Clinical Characterization, Epidemiological and Virological Aspects. PLoS Negl Trop Dis **10**:e0004636.

48 Cerbino-Neto J, Mesquita EC, Souza TM, Parreira V, Wittlin BB, Durovni B, Lemos MC, Vizzoni A, Bispo de Filippis AM, Sampaio SA, Goncalves BS, Bozza FA. 2016. Clinical Manifestations of Zika Virus Infection, Rio de Janeiro, Brazil, 2015. Emerg Infect Dis **22**.

49 Coelho FC, Durovni B, Saraceni V, Lemos C, Codeco CT, Camargo S, de Carvalho LM, Bastos L, Arduini D, Villela DA, Armstrong M. 2016. Higher incidence of Zika in adult women than adult men in Rio de Janeiro suggests a significant contribution of sexual transmission from men to women. Int J Infect Dis **51**:128–132.

50 Cardoso CW, Paploski IA, Kikuti M, Rodrigues MS, Silva MM, Campos GS, Sardi SI, Kitron U, Reis MG, Ribeiro GS. 2015. Outbreak of Exanthematous Illness Associated with Zika, Chikungunya, and Dengue Viruses, Salvador, Brazil. Emerg Infect Dis **21**:2274–2276.

51 Zammarchi L, Tappe D, Fortuna C, Remoli ME, Gunther S, Venturi G, Bartoloni A, Schmidt-Chanasit J. 2015. Zika virus infection in a traveller returning to Europe from Brazil, March 2015. Euro Surveill **20**.

52 Camacho E, Paternina-Gomez M, Blanco PJ, Osorio JE, Aliota MT. 2016. Detection of Autochthonous Zika Virus Transmission in Sincelejo, Colombia. Emerg Infect Dis **22**:927–929.

53 Guerbois M, Fernandez-Salas I, Azar SR, Danis-Lozano R, Alpuche-Aranda CM, Leal G, Garcia-Malo IR, Diaz-Gonzalez EE, Casas-Martinez M, Rossi SL, Del Rio-Galvan SL, Sanchez-Casas RM, Roundy CM, Wood TG, Widen SG, Vasilakis N, Weaver SC. 2016. Outbreak of Zika Virus Infection, Chiapas State, Mexico, 2015, and First Confirmed Transmission by Aedes aegypti Mosquitoes in the Americas. J Infect Dis **214**:1349–1356.

54 Lednicky J, Beau De Rochars VM, El Badry M, Loeb J, Telisma T, Chavannes S, Anilis G, Cella E, Ciccozzi M, Rashid M, Okech B, Salemi M, Morris JG, Jr. 2016. Zika Virus Outbreak in Haiti in 2014: Molecular and Clinical Data. PLoS Negl Trop Dis **10**:e0004687.

55 Journel I, Andrecy LL, Metellus D, Pierre JS, Faublas RM, Juin S, Dismer AM, Fitter DL, Neptune D, Laraque MJ, Corvil S, Pierre M, Buteau J, Lafontant D, Patel R, Lemoine JF, Lowrance DW, Charles M, Boncy J, Adrien P. 2017. Transmission of Zika Virus—Haiti, October 12, 2015–September 10, 2016. MMWR Morb Mortal Wkly Rep **66**:172–176.

56 Adams L, Bello-Pagan M, Lozier M, Ryff KR, Espinet C, Torres J, Perez-Padilla J, Febo MF, Dirlikov E, Martinez A, Munoz-Jordan J, Garcia M, Segarra MO, Malave G, Rivera A, Shapiro-Mendoza C, Rosinger A, Kuehnert MJ, Chung KW, Pate LL, Harris A, Hemme RR, Lenhart A, Aquino G, Zaki S, Read JS, Waterman SH, Alvarado LI, Alvarado-Ramy F, Valencia-Prado M, Thomas D, Sharp TM, Rivera-Garcia B. 2016. Update: Ongoing Zika Virus Transmission—Puerto Rico, November 1, 2015–July 7, 2016. MMWR Morb Mortal Wkly Rep **65**:774–779.

57 Dirlikov E, Ryff KR, Torres-Aponte J, Thomas DL, Perez-Padilla J, Munoz-Jordan J, Caraballo EV, Garcia M, Segarra MO, Malave G, Simeone RM, Shapiro-Mendoza CK, Reyes LR, Alvarado-Ramy F, Harris AF, Rivera A, Major CG, Mayshack M, Alvarado LI, Lenhart A, Valencia-Prado M, Waterman S, Sharp TM, Rivera-Garcia B. 2016. Update: Ongoing Zika Virus Transmission—Puerto Rico, November 1, 2015-April 14, 2016. MMWR Morb Mortal Wkly Rep **65**:451–455.

58 Thomas DL, Sharp TM, Torres J, Armstrong PA, Munoz-Jordan J, Ryff KR, Martinez-Quinones A, Arias-Berrios J, Mayshack M, Garayalde GJ, Saavedra S, Luciano CA, Valencia-Prado M, Waterman S, Rivera-Garcia B. 2016. Local Transmission of Zika Virus—Puerto Rico, November 23, 2015–January 28, 2016. MMWR Morb Mortal Wkly Rep **65**:154–158.

59 De Smet B, Van den Bossche D, van de Werve C, Mairesse J, Schmidt-Chanasit J, Michiels J, Arien KK, Van Esbroeck M, Cnops L. 2016. Confirmed Zika virus infection in a Belgian traveler returning from Guatemala, and the diagnostic challenges of imported cases into Europe. J Clin Virol **80**:8–11.

60 Zambrano H, Waggoner JJ, Almeida C, Rivera L, Benjamin JQ, Pinsky BA. 2016. Zika Virus and Chikungunya Virus CoInfections: A Series of Three Cases from a Single Center in Ecuador. Am J Trop Med Hyg **95**:894–896.

61 Norman FF, Chamorro S, Vazquez A, Sanchez-Seco MP, Perez-Molina JA, Monge-Maillo B, Vivancos MJ, Rodriguez-Dominguez M, Galan JC, de Ory F, Lopez-Velez R. 2016. Sequential Chikungunya and Zika Virus Infections in a Traveler from Honduras. Am J Trop Med Hyg **95**:1166–1168.

62 Piorkowski G, Richard P, Baronti C, Gallian P, Charrel R, Leparc-Goffart I, de Lamballerie X. 2016. Complete coding sequence of Zika virus from Martinique outbreak in 2015. New Microbes New Infect **11**:52–53.

63 Zonneveld R, Roosblad J, Staveren JW, Wilschut JC, Vreden SG, Codrington J. 2016. Three atypical lethal cases associated with acute Zika virus infection in Suriname. IDCases **5**:49–53.

64 Sarmiento-Ospina A, Vasquez-Serna H, Jimenez-Canizales CE, Villamil-Gomez WE, Rodriguez-Morales AJ. 2016. Zika virus associated deaths in Colombia. Lancet Infect Dis **16**:523–524.

65 Broutet N, Krauer F, Riesen M, Khalakdina A, Almiron M, Aldighieri S, Espinal M, Low N, Dye C. 2016. Zika Virus as a Cause of Neurologic Disorders. New England Journal of Medicine **374**:1506–1509.

66 Metsky HC, Matranga CB, Wohl S, Schaffner SF, Freije CA, Winnicki SM, West K, Qu J, Baniecki ML, Gladden-Young A, Lin AE, Tomkins-Tinch CH, Ye SH, Park DJ, Luo CY, Barnes KG, Shah RR, Chak B, Barbosa-Lima G, Delatorre E, Vieira YR, Paul LM, Tan AL, Barcellona CM, Porcelli MC, Vasquez C, Cannons AC, Cone MR, Hogan KN, Kopp EW, Anzinger JJ, Garcia KF, Parham LA, Ramírez RMG, Montoya MCM, Rojas DP, Brown CM, Hennigan S, Sabina B, Scotland S, Gangavarapu K, Grubaugh ND, Oliveira G, Robles-Sikisaka R, Rambaut A, Gehrke L, Smole S, Halloran ME, Villar L, Mattar S, et al. 2017. Zika virus evolution and spread in the Americas. Nature **546**:411–415.

67 Grubaugh ND, Ladner JT, Kraemer MUG, Dudas G, Tan AL, Gangavarapu K, Wiley MR, White S, Thézé J, Magnani DM, Prieto K, Reyes D, Bingham AM, Paul LM, Robles-Sikisaka R, Oliveira G, Pronty D, Barcellona CM, Metsky HC, Baniecki ML, Barnes KG, Chak B, Freije CA, Gladden-Young A, Gnirke A, Luo C, MacInnis B, Matranga CB, Park DJ, Qu J, Schaffner SF, Tomkins-Tinch C, West KL, Winnicki SM, Wohl S, Yozwiak NL, Quick J, Fauver JR, Khan K, Brent SE, Reiner Jr RC, Lichtenberger PN, Ricciardi MJ, Bailey VK, Watkins DI, Cone MR, Kopp Iv EW, Hogan KN, Cannons AC, Jean R, et al. 2017. Genomic epidemiology reveals multiple introductions of Zika virus into the United States. Nature **546**:401–405.

68 Bueno MG, Martinez N, Abdalla L, Duarte Dos Santos CN, Chame M. 2016. Animals in the Zika Virus Life Cycle: What to Expect from Megadiverse Latin American Countries. PLoS Negl Trop Dis **10**:e0005073.

3

ZIKV Transmission and Prevention

3.1 Modes of Transmission

3.1.1 Mosquito-Borne/Bites of Insects

When Zika virus (ZIKV) was first isolated, there was skepticism as to whether it was transmitted by mosquito bite. The virus was isolated from *Aedes africanus* in 1948, but when ZIKV was isolated from a sentinel monkey rhesus (MR766) in 1947, the monkeys were kept caged, preventing possible mosquito bites in these animals. It was only later that the monkeys were kept uncaged on tree platforms to expose them to mosquito bites. The most important question to be answered was whether the mosquitoes were simply passengers or if they could behave as the vector or reservoir of ZIKV, albeit they were only occasionally infected (1, 2).

The first study to demonstrate that an infected mosquito could infect a monkey was by Boorman & Porterfield (1956). They showed that the virus could infect *Aedes aegypti* mosquitoes, and the infection could persist for up to 10 weeks. A monkey was exposed to the infected mosquitoes, and the virus was successfully isolated from the monkey 72 days post-infection (d.p.i.). A high viral load was used for the infection, but it was unknown at that time how much virus a mosquito bite could transmit (3). It is now clear that no right number can be used to mimic an infection in a wild environment. ZIKV strain P6-740 was isolated from *Aedes aegypti* in 1966. This strain was identified in 1 out of 58 pools of 1,277 *Aedes aegypti*, which were collected from cities and towns in Malaya. Interestingly, no ZIKV was isolated from pools

Zika Virus and Diseases: From Molecular Biology to Epidemiology, First Edition.
Suzane Ramos da Silva, Fan Cheng and Shou-Jiang Gao.
© 2018 John Wiley & Sons, Inc. Published 2018 by John Wiley & Sons, Inc.

of 4,492 *Aedes albopictus* captured in suburban and rural areas, or in any other 27,636 *Aedes ssp* collected in different regions of Mayala (4).

After the 2015 outbreak in Brazil, many countries have documented autochthonous transmission of ZIKV (5–10). The first ZIKV infection in an HIV-infected patient was also documented in Brazil. A 38-year-old male patient presented common symptoms of ZIKV infection. He was treated with anti-retroviral therapy, and his CD4$^+$ T cell count was 715 cells/mm^3 at the time of admission with no drastic change for up to 4 weeks after ZIKV infection. ZIKV was detected by reverse transcription-polymerase chain reaction (RT-PCR). There was no infection of dengue virus (DENV) or chikungunya virus (CHIKV) (11).

The circulation of ZIKV infection in *Aedes aegypt* in Brazil was reported in 2016. Mosquitoes were collected indoors or in the surrounding resident areas between June and May 2015 in Rio de Janeiro. From the captured 1,683 mosquitoes, three pools of *Aedes aegypti* (198 pools from of 550 mosquitoes) were positive for ZIKV by RT-PCR, including one with only male mosquitoes, while none of the *Aedes albopictus* (21 pools from 26 mosquitoes) or *Culex quinquefasciatus* (249 pools from 1,107 mosquitoes) was positive (12). The detection of infection in a male mosquito was previously reported in *Aedes furcifer* from Senegal, which might indicate possible vertical and/or venereal transmission of ZIKV in nature (13). Additionally, *Aedes aegypti* was confirmed as the main vector in Brazil since *Culex quinquefasciatus* was not particularly competent to transmit the ZIKV strains circulating in Brazil (14).

Comparison of ZIKV transmission by *Aedes aegypti* and *Aedes albopictus* carried out by collected mosquitoes in different countries and exposed them to the Asian strain from New Caledonia. Both *Aedes aegypti* and *Aedes albopictus* populations exhibited high infection rates and low transmission rates, but *Aedes aegypti* from Guadeloupe and French Guiana had higher viral dissemination rates than the ones isolated in Brazil, the Caribbean Island, and the United States. These findings indicated that other factors, including environment and host, might modulate outbreaks of mosquito-borne pathogens besides the vector (15). One interesting study by Dutra et al. (2016) analyzed *Aedes aegypti* harboring a endosymbiont bacteria named *Wolbachia*, estimated to be in 40.0% of the terrestrial arthropods (16), and compared them with those without bacteria (17). The mosquitoes were exposed to two different isolates of ZIKV from Brazil, and those harboring the bacteria were resistant to ZIKV infection, and lack of dissemination of infection and infectious viral particles in the saliva, suggesting that the transmission was blocked (17).

The transmission cycle of arboviruses such as ZIKV is composed of two parts: the sylvatic and the urban cycle. The sylvatic cycle consists of the transmission of arboviruses from the infected-mosquito to nonhuman primates (such as monkeys), which can then transmit the viruses to noninfected mosquitoes when bitten by them, hence closing the cycle. The urban cycle is the transmission between infected-mosquitoes and humans, and then back to noninfected mosquitoes. The interchange between these cycles occurs when a mosquito first bites a contaminated nonhuman primate, then a noninfected human, or vice versa (Figure 3.1) (18). Each area or country might have a dominant cycle. In Africa, the sylvatic cycle has been reported while in Asia the urban cycle is well-known (19). The establishment of a sylvatic cycle is important to maintain the transmission of the virus, which serves as a reservoir. When this cycle is established, the control of the vector, which transmits the disease might be more complicated. One study analyzed the potential of establishing a sylvatic cycle in

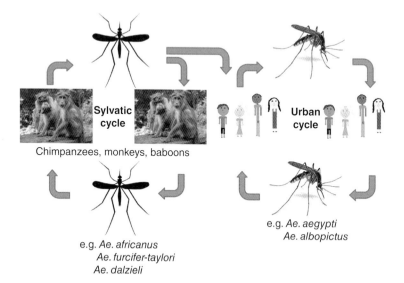

Figure 3.1 Sylvatic and urban cycle involving *Aedes ssp*. Original figure described CHIKV transmission, which also applies to ZIKV and DENV. *Source:* Extracted from Thiboutot et al. (2010) (18).

Brazil, and concluded that there was a high probability of establishing such a cycle in the forests of South America (20).

3.1.2 Vertical and Perinatal Transmission

The first case describing perinatal transmission was during the French Polynesia outbreak (2013–2014). Mother 1 presented symptoms of infection 2 days before the 38-week delivery of a health newborn. ZIKV was detected by RT-PCR in the sera of the mother and newborn, as well as breast milk and mother's saliva, though no infectious virus was found in the breast milk. The second mother had signs of infection, including fever and pruritic rash, 3 days after a 38-week delivery of a newborn that presented severe hypotrophy (intrauterine growth restrictions was detected during pregnancy). The delivery was done by caesarean section. The newborn whom was breastfed starting on day 3 post-delivery developed a skin rash on day 4 after exposition to UV light for jaundice treatment. Both mothers and newborns were positive for ZIKV by RT-PCR on days 2 and 4 after the delivery, respectively. None was positive for DENV. These findings and the timeframe for the infection suggested that the transmission might have occurred trans-placental or at the time of delivery (21).

A study from Tabata et al. (2016) demonstrated that ZIKV infected different cells according to the gestation period of the women. ZIKV infected chorionic villus in early gestation, and primary placental cells in mid- and late-gestations. This indicates that ZIKV might have a placental and a para-placental route of infection (22). This hypothesis is corroborated by the fact that *TIM1*, recently described as one of ZIKV's cellular receptors (23), is expressed in placental villi and amniochorionic membranes (22).

Another study showed that African, Asian, or American strain of ZIKV could infect placenta, which induced anti-inflammatory response. Placenta also expresses *AXL*, one of the main cellular receptors for ZIKV entry (24–27), and ZIKV infection actually induces its expression, pointing to a possible tissue tropism (28). The ability of ZIKV to cross mouse fetal-placenta barrier was demonstrated with Asian SZ01 strain, which was injected by intra-peritoneal (I.P.) into 10-week-old pregnant C57 mice, resulting in abnormal brain development of the offspring (29). Additionally, vertical transmission in *Aedes aegypti*, with a ZIKV strain isolated from Mexico, was documented with a 1:290 rate of infection. Of 69 pools (consisting of 1,738 first adult offspring), 6 were positive for ZIKV, while none of the 32 pools (803 first adult offspring) from *Aedes albopictus* was positive (30).

3.1.3 Sexual Transmission

Sexual transmission of ZIKV has been described in several reports. Since the main route of transmission is by mosquito bite, ZIKV is not classified as a sexually transmitted disease (STD) (31, 32). The first case of sexual transmission of ZIKV was described in a study in 2011, which reported follow-up of three patients in Colorado, United States, in 2008. Patients 1 and 2 were working in the village of Bandafassi, Senegal, and reported being bit by *Aedes ssp.* numerous times when they were abroad. At around 9 days after their return to the United States, both started presenting symptoms, which matched the symptoms described for ZIKV infection in the 1970s and 1980s (33).

Patient 1 was a 36-year-old male who presented symptoms as maculopapular rash, extreme fatigue, headache, aphthous ulcers on the lip, prostatitis, hematospermia, and swelling and arthralgia of wrists, knees, and ankles, but no fever. Patient 2 was a 27-year-old male and presented similar symptoms but no problem with prostate or ulcers. Patient 3 was the first patient's wife, and her symptoms started 9 days after her husband returned. She presented similar symptoms as patients 1 and 2, plus conjunctivitis, and did not have fever. Patients 1 and 2 were vaccinated against YFV before the trip. They were positive for ZIKV and YFV by HI and neutralization tests, while patient 3 was only positive for ZIKV (33).

Additional evidence for ZIKV sexual transmission was based on a report of a 44-year-old male patient from Tahiti, French Polynesia, in December 2013. The patient developed the usual symptoms of ZIKV infection, which went away a few days later but returned in 8 weeks after the first sign. No major symptom was detected after an additional 2 weeks, when the patient developed hematospermia. After admission into the hospital, samples of blood, urine, and semen were collected, which showed no problem in prostate or urinary tract. Three days later, new samples were collected, and tested for ZIKV. Both semen samples and the second urine sample were positive by RT-PCR with viral loads at around 10^7 copies/mL for both semen samples and up to 10^3 copies/mL for the urine. ZIKV infectious particles were detected in the semen samples, suggesting possible sexual transmission of this virus (34).

To test the hypothesis that ZIKV is sexually transmitted, urine, saliva, and blood samples were collected from a couple of a 24-year-old woman and a 44-year-old man. ZIKV was detected in urine (3.5×10^3 copies/mL) and saliva (2.1×10^4 copies/mL) by RT-PCR 3 days after the development of symptoms. Plasma and vaginal swab were negative. The male patient had ZIKV detected in urine at day 16 (4×10^3 copies/mL) and 24 (2.1×10^4 copies/mL) after developing symptoms, and in semen at day 18 (2.9×10^8 copies/mL) and 24 (3.5×10^7 copies/mL). Sequence analysis between the viruses isolated from semen of the male patient and saliva from the female patient only identified four mutations, which were all synonymous—that is, they encode the same amino acids (35).

Even if the incubation time for ZIKV is between 3 and 14 days for symptomatic patients (36), it is challenging to draw a timeline for sexual transmission since late transmission has also been reported. One female patient reported symptoms of ZIKV infection 44 days after her partner had the symptoms, indicating that the transmission might had occurred between 34 and 41 days after the symptoms started in her partner. Both patients (the man aged 61 years and the woman aged 60 years) had traveled to Martinique (a French Caribbean island), but the man was the only one infected abroad, presenting rash, conjunctival hyperemia, arthralgia, and haematospermia without fever. At day 53 after his first symptom (day 9 after her first symptom), ZIKV was not detected in his urine sample by RT-PCR but hers was positive, and by day 67 from his onset of illness, ZIKV detection was negative in semen (37).

The longest documented case for ZIKV persistence in semen was from a patient returning to Italy from Haiti. ZIKV RNA was detected in semen for up to 181 days and in saliva for up to 47 days after the onset of symptoms, but no infectious virus was isolated (38). The same group also reported infectious ZIKV particles isolated from saliva 6 days after the onset of symptoms in a female patient returning to Italy from the Dominican Republic. ZIKV RNA was detected for up to 29 days from both saliva and urine samples. Additional data include a possible sexual transmission from a vasectomized man with detection of infectious ZIKV particles for 69 days in semen (40), as well as detection of ZIKV in spermatozoa (41). Other cases of sexual transmission of ZIKV have been reported in the United States (42, 43) and worldwide such as Colombia (44), France (45), Germany (46), Italy (47), New Zealand (48), and the United Kingdom (49). Sexual transmission is associated with vaginal, anal, and oral sex, and has been reported between men-to-women (50), women-to-men (51), and men-to-men (52).

Detection of ZIKV in the human female genital tract (53, 54) and ZIKV infection of human uterine fibroblasts (55) have been reported. To better understand how the infection might happen in the female reproductive tract, a study was conducted in an animal model, which showed that hormones played an important role. ZIKV was administered into vaginas of AG129 mice (deficient in types I/II interferon receptors) and LysMCre⁺ IFNAR$^{fl/fl}$ C57BL/6 mice (deficient in type I interferon receptor in myeloid cells), which were then exposed to hormonal treatment to mimic different estrus- and diestrus-like phases. Interestingly, AG129 mice in the diestrus-like phase presented lethal infection. For LysMCre⁺ IFNAR$^{fl/fl}$ C57BL/6 mice in the diestrus-like phase infected by ZIKV, viral RNA was detected for up to 10 d.p.i. in vaginal washes. These mice recovered from the illness. Mice at the estrus-like phase were resistant to ZIKV infection. These findings indicate that ZIKV is able to infect and replicate in the female reproductive tract, and most importantly, hormones may influence ZIKV infection (56).

Sexual transmission was also observed in AG129 mice. Persistent infection of ZIKV was detected in testes for up to 4 weeks after infection. In semen, infectious particles were detected from 7 to 21 d.p.i., while ZIKV RNA was detected for up to 5 weeks. Vasectomy prior to the infection drastically decreased ZIKV titers in seminal fluids. Sexual transmission occurred in 50.0% of the mated females (57). A study analyzing the effect of ZIKV infection in testes used wild-type male C57BL/6 mice treated once with a blocking monoclonal antibody of IFN-α and IFN-β receptor 1. Two different ZIKV strains Dakar41519, a mouse-adapted African strain, and H/PF/2013, an Asian one from French Polynesia, were used in the study. As expected, the mouse-adapted strain was more efficient in infection than the Asian one. ZIKV was found in testis and epididymis, mainly in spermatogonia, primary spermatocytes and Sertoli cells in wild-type mice. Interestingly, Axl$^{-/-}$ mice were also positive, indicating that infection of testis was independent of Axl receptor. Decreased levels of testosterone and inhibin B as well as oligospermia were present. Infiltrations of inflammatory cells were found in the testis of the infected mice. Furthermore, Rag1$^{-/-}$ mice (without B and T cells) were used to confirm that adaptive immune response and together with ZIKV infection played a role in causing testis damage. There were also decreased pregnancy and viable fetuses in mice infected with ZIKV (58). Male infertility because of ZIKV infection was suggested based on a study with Ifnar1$^{-/-}$ mice infected by I.P. with ZIKV-SMGC-1 Asian strain. The mice presented orchitis/epididymitis as a result of ZIKV infection in testis and epididymis. Interestingly, ZIKV was not detected in prostate or seminal vesicles, probably because of the lack of Axl expression (59).

3.1.4   ZIKV in the Blood and Other Body Fluids

ZIKV has been detected in blood and other body fluids. During the French Polynesia outbreak, blood samples from donors collected from November 2013 to February 2014 were analyzed, and ZIKV was detected by RT-PCR in 3.0% (42/1,505) of the samples (60). In Martinique, donated blood samples were evaluated between January and June 2016, and ZIKV was detected in 1.84% (76/4,129) of samples. Follow-up of the donors at 7 and 14 days after the donation revealed that 45.3% (34/75) were asymptomatic, while 54.7% (41/75) presented with the classical ZIKV symptoms (61). In Puerto Rico, the detection rate was 0.5% (68/12,777) in the donated blood samples between April and June 2016 (62). On the other hand, in the United States, the positivity rate was 0.001% based on the analysis of 466,834 samples from blood donors from areas where there was no active transmission of ZIKV but had traveled to epidemic places. No ZIKV infection was described in the receivers (63).

The detection and isolation of ZIKV for up to day 7 in conjunctival fluid and for up to 5 days in serum after the onset of symptoms were reported in Chinese travelers who had visited Venezuela. The isolated virus was injected into brains of neonatal BABL/c mice. The virus was detected by RT-PCR, and sequencing analysis revealed that these isolated strains were similar to those isolated in South and Central America (64). Transmission through tears and sweat was also pointed out as a possible route of ZIKV infection when a 38-year-old patient (patient 2) from Utah, United States, was infected after taking care of a 73-year-old patient (patient 1) with ZIKV infection acquired in Mexico. The first patient died 4 days after hospitalization. He presented the highest viral load detected in any patients with ZIKV infection so far with a viral titer of 2.0×10^8 copies/mL.

The second patient who helped with the care without gloves had symptoms 7 days after contact with the first patient. Other health care workers in charge of the first patient had no symptoms (65). The ability of ZIKV to infect retina and other eye regions was showed by Miner et al. (2016). Ifnar$1^{-/-}$ mice or wild-type mice treated with anti-Ifnar1 blocking antibody were infected with Asian strains (H/PF/2013 or Paraiba 2015) by subcutaneous injection (S.C.) on the footpad. The animals developed conjunctivitis and panuveitis. Animals with Axl and/or Mertk knockout also maintained comparable levels of infection, indicating a route of infection independent of these receptors (66).

ZIKV transmitted by platelet transfusion was reported in two recipients in Rio de Janeiro, Brazil. The donor was asymptomatic at the time of donation but contacted the hospital 2 days after describing symptoms associated with ZIKV infection. Recipient 1 was a 54-year-old female patient with primary myelofibrosis syndrome, while recipient 2 was a 14-year-old female patient with acute myeloid leukemia with immunosuppression at the time of a haploid-identical bone-marrow transplant. The serum of the donor obtained 3 days after the symptoms was tested positive for ZIKV by RT-PCR and plaque-reduction neutralization testing (PRNT). The serum from the first recipient was tested positive by RT-PCR at day 6 and by PRNT at day 31 after transfusion. The second recipient had ZIKV detected in her serum by RT-PCR and PRNT at day 23 and 51 after the transfusion. None of the recipients developed symptoms related with ZIKV infection (67). Additionally, one case of occupational ZIKV infection with classical symptoms has been reported but without serious consequence (68).

3.2 Prevention

The Centers for Disease Control and Prevention (CDC) has developed guidelines to avoid ZIKV infection, especially for pregnant women, or women who are planning a pregnancy.

3.2.1 Avoiding Sexual Transmission of ZIKV

For couples with one of the partners who has lived or traveled to endemic areas, it is recommended that condoms should be used for at least six months if they do not want to abstain from sex. In the case of pregnant women, the use of condoms is recommended until the end of the pregnancy. Male or female condoms should be used in any sexual act, which includes vaginal, anal, and oral sex. For women who are planning pregnancy, it is recommended to wait for at least 6 months to conceive if the partner has traveled to endemic areas regardless of his symptoms. There is no information available regarding problems with future pregnancy for women pre-exposed to ZIKV infection (69–71).

3.2.2 Avoiding Mosquito Bites

Covering arms and legs is the basic recommendation; however, ZIKV is spreading fast in countries with a tropical climate, which means people will probably not wear long sleeves and pants. The basic protocols are to avoid forested areas that are known to have mosquitoes, and to use mosquito-repellent in the exposed areas of skin, especially if the products contain DEET or p-menthane-3,8-diol, which has been shown to be protective for an extended period of time (72). Doors and windows should have screens to prevent mosquitoes to get inside the house. The use of mosquito-netting over beds and cribs is also recommended. *Aedes aegypti* are active for most of day, meaning they have feeding habits that are different from other mosquitoes. Their feeding time is not restricted to sunset; and they might bite at night as well (73–76).

The best way to avoid infection in urban and suburban areas is to decrease the number of mosquitoes. The mosquitoes deposit their eggs in not-moving water regardless if it is clean or not. The biggest challenge in cities is to eliminate water-filled containers in the backyards of the houses (Figure 3.2) (77). These are the favorite places for mosquitoes to deposit their eggs. Empty containers should be kept upside-down or covered, and dishes under flower vases should have little holes to avoid retention of water (Figure 3.3) (78). Water containers for pets should be cleaned and kept with fresh water. When throwing away water from containers, it is necessary to wash its inside walls, especially in regions where water was stagnated. Mosquitoes deposit their eggs on to the walls of the containers if they contain water. These eggs should be removed by scratching the walls of the containers. If not, once fresh water is added, the eggs will further develop (79, 80).

Figure 3.2 Example of a backyard with containers that will accumulate water in the rainy season, promoting deposition of *Aedes ssp.* eggs in the water.
Source: Extracted from Ledermann et al. (2014) (77).

Figure 3.3 Water accumulated in the plate under the flower vase. This is one of the most common places *Aedes ssp.* deposit their eggs, contributing to ZIKV dissemination. *Source:* Extracted from Reiter (2016) (78).

References

1 Haddow AJ, Williams MC, Woodall JP, Simpson DI, Goma LK. 1964. Twelve Isolations of Zika Virus from Aedes (Stegomyia) Africanus (Theobald) Taken in and above a Uganda Forest. Bull World Health Organ **31**:57–69.
2 Mattingly PF. 1971. Ecological aspects of mosquito evolution. Parassitologia **13**:31–65.
3 Boorman JP, Porterfield JS. 1956. A simple technique for infection of mosquitoes with viruses; transmission of Zika virus. Trans R Soc Trop Med Hyg **50**:238–242.
4 Marchette NJ, Garcia R, Rudnick A. 1969. Isolation of Zika virus from Aedes aegypti mosquitoes in Malaysia. Am J Trop Med Hyg **18**:411–415.
5 Likos A, Griffin I, Bingham AM, Stanek D, Fischer M, White S, Hamilton J, Eisenstein L, Atrubin D, Mulay P, Scott B, Jenkins P, Fernandez D, Rico E, Gillis L, Jean R, Cone M, Blackmore C, McAllister J, Vasquez C, Rivera L, Philip C. 2016. Local Mosquito-Borne Transmission of Zika Virus - Miami-Dade and Broward Counties, Florida, June-August 2016. MMWR Morb Mortal Wkly Rep **65**:1032–1038.

6 Zanluca C, Melo VC, Mosimann AL, Santos GI, Santos CN, Luz K. 2015. First report of autochthonous transmission of Zika virus in Brazil. Mem Inst Oswaldo Cruz **110**:569–572.

7 Camacho E, Paternina-Gomez M, Blanco PJ, Osorio JE, Aliota MT. 2016. Detection of Autochthonous Zika Virus Transmission in Sincelejo, Colombia. Emerg Infect Dis **22**:927–929.

8 Guerbois M, Fernandez-Salas I, Azar SR, Danis-Lozano R, Alpuche-Aranda CM, Leal G, Garcia-Malo IR, Diaz-Gonzalez EE, Casas-Martinez M, Rossi SL, Del Rio-Galvan SL, Sanchez-Casas RM, Roundy CM, Wood TG, Widen SG, Vasilakis N, Weaver SC. 2016. Outbreak of Zika Virus Infection, Chiapas State, Mexico, 2015, and First Confirmed Transmission by Aedes aegypti Mosquitoes in the Americas. J Infect Dis **214**:1349–1356.

9 Jimenez Corona ME, De la Garza Barroso AL, Rodriguez Martinez JC, Luna Guzman NI, Ruiz Matus C, Diaz Quinonez JA, Lopez Martinez I, Kuri Morales PA. 2016. Clinical and Epidemiological Characterization of Laboratory-Confirmed Autochthonous Cases of Zika Virus Disease in Mexico. PLoS Curr **8**.

10 Hall-Mendelin S, Pyke AT, Moore PR, Mackay IM, McMahon JL, Ritchie SA, Taylor CT, Moore FAJ, van den Hurk AF. 2016. Assessment of Local Mosquito Species Incriminates Aedes aegypti as the Potential Vector of Zika Virus in Australia. PLOS Neglected Tropical Diseases **10**:e0004959.

11 Calvet GA, Filippis AM, Mendonca MC, Sequeira PC, Siqueira AM, Veloso VG, Nogueira RM, Brasil P. 2016. First detection of autochthonous Zika virus transmission in a HIV-infected patient in Rio de Janeiro, Brazil. J Clin Virol **74**:1–3.

12 Ferreira-de-Brito A, Ribeiro IP, Miranda RM, Fernandes RS, Campos SS, Silva KA, Castro MG, Bonaldo MC, Brasil P, Lourenco-de-Oliveira R. 2016. First detection of natural infection of Aedes aegypti with Zika virus in Brazil and throughout South America. Mem Inst Oswaldo Cruz **111**:655–658.

13 Diallo D, Sall AA, Diagne CT, Faye O, Faye O, Ba Y, Hanley KA, Buenemann M, Weaver SC, Diallo M. 2014. Zika virus emergence in mosquitoes in southeastern Senegal, 2011. PLoS One **9**:e109442.

14 Fernandes RS, Campos SS, Ferreira-de-Brito A, Miranda RM, Barbosa da Silva KA, Castro MG, Raphael LM, Brasil P, Failloux AB, Bonaldo MC, Lourenco-de-Oliveira R. 2016. Culex quinquefasciatus from Rio de Janeiro Is Not Competent to Transmit the Local Zika Virus. PLoS Negl Trop Dis **10**:e0004993.

15 Chouin-Carneiro T, Vega-Rua A, Vazeille M, Yebakima A, Girod R, Goindin D, Dupont-Rouzeyrol M, Lourenco-de-Oliveira R, Failloux AB. 2016. Differential Susceptibilities of Aedes aegypti and Aedes albopictus from the Americas to Zika Virus. PLoS Negl Trop Dis **10**:e0004543.

16 Zug R, Hammerstein P. 2012. Still a host of hosts for Wolbachia: analysis of recent data suggests that 40% of terrestrial arthropod species are infected. PLoS One 7:e38544.

17 Dutra HL, Rocha MN, Dias FB, Mansur SB, Caragata EP, Moreira LA. 2016. Wolbachia Blocks Currently Circulating Zika Virus Isolates in Brazilian Aedes aegypti Mosquitoes. Cell Host Microbe **19**:771–774.

18 Thiboutot MM, Kannan S, Kawalekar OU, Shedlock DJ, Khan AS, Sarangan G, Srikanth P, Weiner DB, Muthumani K. 2010. Chikungunya: a potentially emerging epidemic? PLoS Negl Trop Dis 4:e623.

19 Petersen LR, Jamieson DJ, Powers AM, Honein MA. 2016. Zika Virus. New England Journal of Medicine **374**:1552–1563.

20 Althouse BM, Vasilakis N, Sall AA, Diallo M, Weaver SC, Hanley KA. 2016. Potential for Zika Virus to Establish a Sylvatic Transmission Cycle in the Americas. PLoS Negl Trop Dis **10**:e0005055.

21 Besnard M, Lastere S, Teissier A, Cao-Lormeau V, Musso D. 2014. Evidence of perinatal transmission of Zika virus, French Polynesia, December 2013 and February 2014. Euro Surveill **19**.

22 Tabata T, Petitt M, Puerta-Guardo H, Michlmayr D, Wang C, Fang-Hoover J, Harris E, Pereira L. 2016. Zika Virus Targets Different Primary Human Placental Cells, Suggesting Two Routes for Vertical Transmission. Cell Host Microbe **20**:155–166.

23 Hamel R, Dejarnac O, Wichit S, Ekchariyawat P, Neyret A, Luplertlop N, Perera-Lecoin M, Surasombatpattana P, Talignani L, Thomas F, Cao-Lormeau VM, Choumet V, Briant L, Despres P, Amara A, Yssel H, Misse D. 2015. Biology of Zika Virus Infection in Human Skin Cells. J Virol **89**:8880–8896.

24 Liu S, DeLalio LJ, Isakson BE, Wang TT. 2016. AXL-Mediated Productive Infection of Human Endothelial Cells by Zika Virus. Circ Res **119**:1183–1189.

25 Meertens L, Labeau A, Dejarnac O, Cipriani S, Sinigaglia L, Bonnet-Madin L, Le Charpentier T, Hafirassou ML, Zamborlini A, Cao-Lormeau VM, Coulpier M, Misse D, Jouvenet N, Tabibiazar R, Gressens P, Schwartz O, Amara A. 2017. Axl Mediates ZIKA Virus Entry in Human Glial Cells and Modulates Innate Immune Responses. Cell Rep **18**:324–333.

26 Richard AS, Shim BS, Kwon YC, Zhang R, Otsuka Y, Schmitt K, Berri F, Diamond MS, Choe H. 2017. AXL-dependent infection of human fetal endothelial cells distinguishes Zika virus from other pathogenic flaviviruses. Proc Natl Acad Sci U S A **114**:2024–2029.

27 Nowakowski Tomasz J, Pollen Alex A, Di Lullo E, Sandoval-Espinosa C, Bershteyn M, Kriegstein Arnold R. Expression Analysis Highlights AXL as a Candidate Zika Virus Entry Receptor in Neural Stem Cells. Cell Stem Cell **18**:591–596.

28 Vermillion MS, Lei J, Shabi Y, Baxter VK, Crilly NP, McLane M, Griffin DE, Pekosz A, Klein SL, Burd I. 2017. Intrauterine Zika virus infection of pregnant immunocompetent mice models transplacental transmission and adverse perinatal outcomes. Nat Commun **8**:14575.

29 Wu KY, Zuo GL, Li XF, Ye Q, Deng YQ, Huang XY, Cao WC, Qin CF, Luo ZG. 2016. Vertical transmission of Zika virus targeting the radial glial cells affects cortex development of offspring mice. Cell Res **26**:645–654.

30 Thangamani S, Huang J, Hart CE, Guzman H, Tesh RB. 2016. Vertical Transmission of Zika Virus in Aedes aegypti Mosquitoes. Am J Trop Med Hyg **95**:1169–1173.

31 Althaus CL, Low N. 2016. How Relevant Is Sexual Transmission of Zika Virus? PLoS Med **13**:e1002157.

32 Yakob L, Kucharski A, Hue S, Edmunds WJ. 2016. Low risk of a sexually-transmitted Zika virus outbreak. Lancet Infect Dis **16**:1100–1102.

33 Foy BD, Kobylinski KC, Chilson Foy JL, Blitvich BJ, Travassos da Rosa A, Haddow AD, Lanciotti RS, Tesh RB. 2011. Probable non-vector-borne transmission of Zika virus, Colorado, USA. Emerg Infect Dis **17**:880–882.

34 Musso D, Roche C, Robin E, Nhan T, Teissier A, Cao-Lormeau VM. 2015. Potential sexual transmission of Zika virus. Emerg Infect Dis **21**:359–361.

35 D'Ortenzio E, Matheron S, Yazdanpanah Y, de Lamballerie X, Hubert B, Piorkowski G, Maquart M, Descamps D, Damond F, Leparc-Goffart I. 2016. Evidence of Sexual Transmission of Zika Virus. N Engl J Med **374**:2195–2198.

36 Krow-Lucal ER, Biggerstaff BJ, Staples JE. 2017. Estimated Incubation Period for Zika Virus Disease. Emerg Infect Dis **23**.

37 Turmel JM, Abgueguen P, Hubert B, Vandamme YM, Maquart M, Le Guillou-Guillemette H, Leparc-Goffart I. 2016. Late sexual transmission of Zika virus related to persistence in the semen. Lancet **387**:2501.

38 Barzon L, Pacenti M, Franchin E, Lavezzo E, Trevisan M, Sgarabotto D, Palu G. 2016. Infection dynamics in a traveller with persistent shedding of Zika virus RNA in semen for six months after returning from Haiti to Italy, January 2016. Euro Surveill **21**.

39 Barzon L, Pacenti M, Berto A, Sinigaglia A, Franchin E, Lavezzo E, Brugnaro P, Palu G. 2016. Isolation of infectious Zika virus from saliva and prolonged viral RNA shedding in a traveller returning from the Dominican Republic to Italy, January 2016. Euro Surveill **21**.

40 Arsuaga M, Bujalance SG, Diaz-Menendez M, Vazquez A, Arribas JR. 2016. Probable sexual transmission of Zika virus from a vasectomised man. Lancet Infect Dis **16**:1107.

41 Mansuy JM, Suberbielle E, Chapuy-Regaud S, Mengelle C, Bujan L, Marchou B, Delobel P, Gonzalez-Dunia D, Malnou CE, Izopet J, Martin-Blondel G. 2016. Zika virus in semen and spermatozoa. Lancet Infect Dis **16**:1106–1107.

42 Brooks RB, Carlos MP, Myers RA, White MG, Bobo-Lenoci T, Aplan D, Blythe D, Feldman KA. 2016. Likely Sexual Transmission of Zika Virus from a Man with No Symptoms of Infection - Maryland, 2016. MMWR Morb Mortal Wkly Rep **65**:915–916.

43 Russell K, Hills SL, Oster AM, Porse CC, Danyluk G, Cone M, Brooks R, Scotland S, Schiffman E, Fredette C, White JL, Ellingson K, Hubbard A, Cohn A, Fischer M, Mead P, Powers AM, Brooks JT. 2017. Male-to-Female Sexual Transmission of Zika Virus-United States, January-April 2016. Clin Infect Dis **64**:211–213.

44 Towers S, Brauer F, Castillo-Chavez C, Falconar AK, Mubayi A, Romero-Vivas CM. 2016. Estimate of the reproduction number of the 2015 Zika virus outbreak in Barranquilla, Colombia, and estimation of the relative role of sexual transmission. Epidemics **17**:50–55.

45 Freour T, Mirallie S, Hubert B, Splingart C, Barriere P, Maquart M, Leparc-Goffart I. 2016. Sexual transmission of Zika virus in an entirely asymptomatic couple returning from a Zika epidemic area, France, April 2016. Euro Surveill **21**.

46 Frank C, Cadar D, Schlaphof A, Neddersen N, Gunther S, Schmidt-Chanasit J, Tappe D. 2016. Sexual transmission of Zika virus in Germany, April 2016. Euro Surveill **21**.

47 Venturi G, Zammarchi L, Fortuna C, Remoli ME, Benedetti E, Fiorentini C, Trotta M, Rizzo C, Mantella A, Rezza G, Bartoloni A. 2016.

An autochthonous case of Zika due to possible sexual transmission, Florence, Italy, 2014. Euro Surveill **21**.

48 Harrower J, Kiedrzynski T, Baker S, Upton A, Rahnama F, Sherwood J, Huang QS, Todd A, Pulford D. 2016. Sexual Transmission of Zika Virus and Persistence in Semen, New Zealand, 2016. Emerg Infect Dis **22**:1855–1857.

49 Gulland A. 2016. First case of Zika virus spread through sexual contact is detected in UK. BMJ **355**:i6500.

50 Coelho FC, Durovni B, Saraceni V, Lemos C, Codeco CT, Camargo S, de Carvalho LM, Bastos L, Arduini D, Villela DA, Armstrong M. 2016. Higher incidence of Zika in adult women than adult men in Rio de Janeiro suggests a significant contribution of sexual transmission from men to women. Int J Infect Dis **51**:128–132.

51 Davidson A, Slavinski S, Komoto K, Rakeman J, Weiss D. 2016. Suspected Female-to-Male Sexual Transmission of Zika Virus - New York City, 2016. MMWR Morb Mortal Wkly Rep **65**:716–717.

52 Deckard DT, Chung WM, Brooks JT, Smith JC, Woldai S, Hennessey M, Kwit N, Mead P. 2016. Male-to-Male Sexual Transmission of Zika Virus–Texas, January 2016. MMWR Morb Mortal Wkly Rep **65**:372–374.

53 Prisant N, Bujan L, Benichou H, Hayot PH, Pavili L, Lurel S, Herrmann C, Janky E, Joguet G. 2016. Zika virus in the female genital tract. Lancet Infect Dis **16**:1000–1001.

54 Visseaux B, Mortier E, Houhou-Fidouh N, Brichler S, Collin G, Larrouy L, Charpentier C, Descamps D. 2016. Zika virus in the female genital tract. Lancet Infect Dis **16**:1220.

55 Chen JC, Wang Z, Huang H, Weitz SH, Wang A, Qiu X, Baumeister MA, Uzgiris A. 2016. Infection of human uterine fibroblasts by Zika virus in vitro: implications for viral transmission in women. Int J Infect Dis **51**:139–140.

56 Tang WW, Young MP, Mamidi A, Regla-Nava JA, Kim K, Shresta S. 2016. A Mouse Model of Zika Virus Sexual Transmission and Vaginal Viral Replication. Cell Rep **17**:3091–3098.

57 Duggal NK, Ritter JM, Pestorius SE, Zaki SR, Davis BS, Chang GJ, Bowen RA, Brault AC. 2017. Frequent Zika Virus Sexual Transmission and Prolonged Viral RNA Shedding in an Immunodeficient Mouse Model. Cell Rep **18**:1751–1760.

58 Govero J, Esakky P, Scheaffer SM, Fernandez E, Drury A, Platt DJ, Gorman MJ, Richner JM, Caine EA, Salazar V, Moley KH,

Diamond MS. 2016. Zika virus infection damages the testes in mice. Nature **540**:438–442.

59 Ma W, Li S, Ma S, Jia L, Zhang F, Zhang Y, Zhang J, Wong G, Zhang S, Lu X, Liu M, Yan J, Li W, Qin C, Han D, Qin C, Wang N, Li X, Gao GF. 2016. Zika Virus Causes Testis Damage and Leads to Male Infertility in Mice. Cell **167**:1511–1524 e1510.

60 Musso D, Nhan T, Robin E, Roche C, Bierlaire D, Zisou K, Shan Yan A, Cao-Lormeau VM, Broult J. 2014. Potential for Zika virus transmission through blood transfusion demonstrated during an outbreak in French Polynesia, November 2013 to February 2014. Euro Surveill **19**.

61 Gallian P, Cabie A, Richard P, Paturel L, Charrel RN, Pastorino B, Leparc-Goffart I, Tiberghien P, de Lamballerie X. 2017. Zika virus in asymptomatic blood donors in Martinique. Blood **129**:263–266.

62 Kuehnert MJ, Basavaraju SV, Moseley RR, Pate LL, Galel SA, Williamson PC, Busch MP, Alsina JO, Climent-Peris C, Marks PW, Epstein JS, Nakhasi HL, Hobson JP, Leiby DA, Akolkar PN, Petersen LR, Rivera-Garcia B. 2016. Screening of Blood Donations for Zika Virus Infection - Puerto Rico, April 3-June 11, 2016. MMWR Morb Mortal Wkly Rep **65**:627–628.

63 Williamson PC, Linnen JM, Kessler DA, Shaz BH, Kamel H, Vassallo RR, Winkelman V, Gao K, Ziermann R, Menezes J, Thomas S, Holmberg JA, Bakkour S, Stone M, Lu K, Simmons G, Busch MP. 2017. First cases of Zika virus-infected US blood donors outside states with areas of active transmission. Transfusion doi:10.1111/trf.14041.

64 Sun J, Wu, Zhong H, Guan D, Zhang H, Tan Q, Ke C. 2016. Presence of Zika Virus in Conjunctival Fluid. JAMA Ophthalmol doi:10.1001/jamaophthalmol.2016.3417.

65 Swaminathan S, Schlaberg R, Lewis J, Hanson KE, Couturier MR. 2016. Fatal Zika Virus Infection with Secondary Nonsexual Transmission. N Engl J Med doi:10.1056/NEJMc1610613.

66 Miner JJ, Sene A, Richner JM, Smith AM, Santeford A, Ban N, Weger-Lucarelli J, Manzella F, Ruckert C, Govero J, Noguchi KK, Ebel GD, Diamond MS, Apte RS. 2016. Zika Virus Infection in Mice Causes Panuveitis with Shedding of Virus in Tears. Cell Rep **16**:3208–3218.

67 Motta IJ, Spencer BR, Cordeiro da Silva SG, Arruda MB, Dobbin JA, Gonzaga YB, Arcuri IP, Tavares RC, Atta EH, Fernandes RF, Costa DA, Ribeiro LJ, Limonte F, Higa LM, Voloch CM, Brindeiro RM, Tanuri A, Ferreira OC, Jr. 2016. Evidence for Transmission of Zika Virus by Platelet Transfusion. N Engl J Med **375**:1101–1103.

68 Olson CK, Iwamoto M, Perkins KM, Polen KN, Hageman J, Meaney-Delman D, Igbinosa, II, Khan S, Honein MA, Bell M, Rasmussen SA, Jamieson DJ. 2016. Preventing Transmission of Zika Virus in Labor and Delivery Settings Through Implementation of Standard Precautions - United States, 2016. MMWR Morb Mortal Wkly Rep **65**:290–292.

69 Oster AM, Russell K, Stryker JE, Friedman A, Kachur RE, Petersen EE, Jamieson DJ, Cohn AC, Brooks JT. 2016. Update: Interim Guidance for Prevention of Sexual Transmission of Zika Virus–United States, 2016. MMWR Morb Mortal Wkly Rep **65**:323–325.

70 Brooks JT, Friedman A, Kachur RE, LaFlam M, Peters PJ, Jamieson DJ. 2016. Update: Interim Guidance for Prevention of Sexual Transmission of Zika Virus - United States, July 2016. MMWR Morb Mortal Wkly Rep **65**:745–747.

71 Petersen EE, Meaney-Delman D, Neblett-Fanfair R, Havers F, Oduyebo T, Hills SL, Rabe IB, Lambert A, Abercrombie J, Martin SW, Gould CV, Oussayef N, Polen KN, Kuehnert MJ, Pillai SK, Petersen LR, Honein MA, Jamieson DJ, Brooks JT. 2016. Update: Interim Guidance for Preconception Counseling and Prevention of Sexual Transmission of Zika Virus for Persons with Possible Zika Virus Exposure - United States, September 2016. MMWR Morb Mortal Wkly Rep **65**:1077–1081.

72 Rodriguez SD, Drake LL, Price DP, Hammond JI, Hansen IA. 2015. The Efficacy of Some Commercially Available Insect Repellents for Aedes aegypti (Diptera: Culicidae) and Aedes albopictus (Diptera: Culicidae). J Insect Sci **15**:140.

73 Chagas AC, Ramirez JL, Jasinskiene N, James AA, Ribeiro JM, Marinotti O, Calvo E. 2014. Collagen-binding protein, Aegyptin, regulates probing time and blood feeding success in the dengue vector mosquito, Aedes aegypti. Proc Natl Acad Sci U S A **111**:6946–6951.

74 Ribeiro JM. 2000. Blood-feeding in mosquitoes: probing time and salivary gland anti-haemostatic activities in representatives of three genera (Aedes, Anopheles, Culex). Med Vet Entomol **14**:142–148.

75 Liebman KA, Stoddard ST, Reiner RC, Jr., Perkins TA, Astete H, Sihuincha M, Halsey ES, Kochel TJ, Morrison AC, Scott TW. 2014. Determinants of Heterogeneous Blood Feeding Patterns by Aedes aegypti in Iquitos, Peru. PLOS Neglected Tropical Diseases **8**:e2702.

76 Ponlawat A, Harrington LC. 2005. Blood feeding patterns of Aedes aegypti and Aedes albopictus in Thailand. J Med Entomol **42**:844–849.

77 Ledermann JP, Guillaumot L, Yug L, Saweyog SC, Tided M, Machieng P, Pretrick M, Marfel M, Griggs A, Bel M, Duffy MR, Hancock WT, Ho-Chen T, Powers AM. 2014. Aedes hensilli as a Potential Vector of Chikungunya and Zika Viruses. PLoS Neglected Tropical Diseases **8**:e3188.

78 Reiter P. 2016. Control of Urban Zika Vectors: Should We Return to the Successful PAHO/WHO Strategy? PLoS Negl Trop Dis **10**:e0004769.

79 Sharma A, Lal SK. 2017. Zika Virus: Transmission, Detection, Control, and Prevention. Frontiers in Microbiology **8**:110.

80 Shuaib W, Stanazai H, Abazid AG, Mattar AA. 2016. Re-Emergence of Zika Virus: A Review on Pathogenesis, Clinical Manifestations, Diagnosis, Treatment, and Prevention. The American Journal of Medicine **129**:879.e877–879.e812.

4

Association with Guillain-Barré Syndrome and Microcephaly

4.1   Association with Neurological Disorders

Recently, Zika virus (ZIKV) infection has been associated with neurological disorders in adults and newborns. Guillain-Barré syndrome (1–6) and acute disseminated encephalomyelitis (ADEM) have been described in adults (7), while microcephaly has been reported in newborns (8–10). None of these symptoms had been described before the outbreak in French Polynesia in 2013 (11). An overwhelming number of studies have attempted to understand the differences between the virus strains, the association of the different strains with the associated-diseases, and the reproducibility of the same diseases in animal models. ZIKV was fatal in some cases, mostly associated with immunocompromised patients (12, 13). Additionally, the link between ZIKV infection and adult encephalitis and myelitis is still unclear (14).

4.1.1   Guillain-Barré Syndrome

Guillain-Barré syndrome (GBS) is an immune-mediated flaccid paralysis. The first description of this disease was controversial. In 1859, Jean Baptiste Octave Landry reported an "acute ascending paralysis," which was called "Landry's ascending paralysis," without describing the cause (15). In 1916, Georges Guillain, Jean Alexandre Barré, and André Strohl described an acute flaccid paralysis that was characterized by high protein levels in the cerebrospinal fluid (CFS) without increase in inflammatory cells (16). The acute paralysis was previously associated with polio and syphilis infection, and was characterized by

Zika Virus and Diseases: From Molecular Biology to Epidemiology, First Edition.
Suzane Ramos da Silva, Fan Cheng and Shou-Jiang Gao.
© 2018 John Wiley & Sons, Inc. Published 2018 by John Wiley & Sons, Inc.

an increase in white cells. The syndrome was first named Landry–Guillain–Barré–Strohl. However, they believed the disease was not the same as Landry's, so the syndrome was then renamed Guillain-Barré syndrome. The first symptoms are characterized by weakness and a tingling sensation in the extremities, followed by quick paralysis of the whole body. The paralysis is usually transient, and most of the people recover in 6 months or 1 year, but some patients might take many years to recover and are chronically affected by pain. The current treatment is plasmapheresis and immunoglobulin therapy. Although the exactly cause is still unknown, it appears to be associated with many bacterial and viral infections (17, 18).

After the ZIKV outbreak in French Polynesia in 2013, there was an increase of GBS cases. One case of GBS associated with ZIKV was described in a woman in her early 40s. She was diagnosed with GBS 7 days after presenting ZIKV infection symptoms, such as myalgia, slight fever, rash, and conjunctivitis. Serological results indicated recent exposure to ZIKV and previous exposure to Dengue virus (DENV) types 1–4. The patient recovered well and was able to walk independently 40 days after GBS diagnosis (1).

Prior to the ZIKV outbreak, French Polynesia had outbreaks of DENV types 1 and 3 but there was an abrupt increase of GBS 4 weeks after the ZIKV outbreak, indicating their possible association. Of 42 cases of GBS analyzed from November 2013 to April 2014, 35 patients had symptoms associated with ZIKV infection in the weeks prior to GBS diagnosis. Although it was difficult to assess the role of each virus in GBS development, it was interesting to observe an association between the timing of GBS development and ZIKV outbreak (4).

To investigate the role of each virus in GBS, the samples were further analyzed. Of the 42 samples, 39 (93.0%) had IgM antibodies and 42 (100%) had neutralizing antibodies against ZIKV. Furthermore, 37 of the 42 subjects (88.0%) had symptoms compatible with ZIKV infection 4 to 10 days before GBS signs. Since there was no difference of DENV infection between the previous group and this group, it was concluded that GBS was more likely to be caused by ZIKV (19).

A possible association between GBS and ZIKV was also described during the 2015 outbreak of ZIKV in Brazil (20–22). Since then, other countries in Central and South America, such as Colombia (23, 24), Martinique (5), Suriname (25), Guyana (26), Haiti (27), Puerto Rico (28), and others (29) have also observed an association of GBS development with ZIKV infection. Although the timing of ZIKV infection and GBS development is convincing, there is still doubt about the likelihood of a causal role of ZIKV in GBS (3, 30, 31).

4.1.2 Microcephaly

After the 2015 ZIKV outbreak in Brazil, an abrupt increase of microcephaly cases in newborns was described. Microcephaly is a postnatal definition of a severe decrease in the head circumference—that is, a newborn with a reduction of head circumference by about two standard deviations (SD) or with a head circumference less than the third percentile in a severe case compared with other normal babies of the same age and sex (32). In Brazil, the cut-off for term newborns diagnosed with microcephaly was set at 32 cm in December 2015 (33) (Figure 4.1). Many infectious agents have been associated with microcephaly. They are named *TORCH*, which includes toxoplasmosis, other infections, rubella, cytomegalovirus (CMV), and herpes simplex virus (HSV), whereas others may include syphilis, HIV, Varicella-Zoster Virus (VZV), parvovirus B19, and enterovirus (34). Since the ZIKV outbreak in Brazil, this virus has been suggested as another causal agent for microcephaly (35). Although the number of initial reported cases of microcephaly has been revised, and the new criteria include molecular detection of ZIKV, the final number of confirmed cases between the end

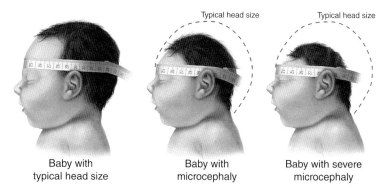

| Baby with typical head size | Baby with microcephaly | Baby with severe microcephaly |

Figure 4.1 Illustration of the differences in head circumferences between microcephaly cases and the normal newborns. On December 2015, Brazil classified a newborn as having microcephaly if the head circumference is under or equal to 32 cm. *Source:* Figure and legend extracted from Centers for Disease Control and Prevention, National Center on Birth Defects and Developmental Disabilities.

of 2015 and the beginning of 2016 remains at least five times higher than those annually reported prior to 2015 (33).

Mlakar et al. (2016) reported the first case of ZIKV detection in the brain tissue of a fetus with microcephaly from a 25-year-old pregnant woman who was infected in Rio Grande do Norte state (Northeast of Brazil) at probably 13 gestational weeks. She returned to Slovenia at 28 weeks. At 29 weeks, the fetus showed some signs of abnormalities, and at 32 weeks, an intrauterine growth retardation (IGR) and calcification in the placenta and brain was detected. The gestation was terminated. The autopsy indicated that the head circumference was 26 cm without other exterior abnormality. ZIKV RNA was detected only in the brain and not in other organs. Other infectious agents related with microcephaly were not detected. Many abnormalities were found in the brain, including calcifications and loss of cortex gyros, sylvian fissures abnormally opened, and alterations in the ventricle. The ZIKV strain isolated from the brain was 99.7% identical to the strain isolated from French Polynesia in 2013 (36).

Two cases from Paraiba state in the Northeast of Brazil were reported. A 27-year-old woman presented ZIKV infection symptoms at 18 weeks of pregnancy. At 21 weeks, ultrasonogram detected some abnormalities in the fetus brain, which were confirmed at 27 weeks. The newborn was delivered at 40 weeks with a 30 cm head circumference. The second woman was 35 years old with an infection probably occurring at 10 gestational weeks. At week 22 and 25, ultrasonogram detected irregularities in the fetus brain, and the neonate presented many development problems. The virus was detected at week 28 in the amniotic fluid but not the serum or urine. DENV and chikungunya (CHIKV) virus as well as other infectious agents associated with microcephaly were not detected. Sequence analysis showed that the virus was closely related to the strain isolated in French Polynesia in 2013 (37).

ZIKV was also isolated from a fetus brain, whose mother was probably infected at week 11 of pregnancy, in November 2015 in Guatemala. A decrease in the head circumference was detected between week 16 and 20. ZIKV RNA was detected at week 16 by reverse-transcript polymerase chain reaction (RT-PCR), indicating a prolonged maternal viremia. Ultrasonogram at week 19 and magnetic resonance imaging (MRI) at week 20 revealed significant brain development alterations, mainly in frontal and parietal lobes, and the pregnancy was terminated at week 21. ZIKV was detected in many fetal organs but the highest viral load was detected in the brain. Other infectious agents were not identified. The virus sequence was similar to other strains isolated in Guatemala (38).

The association between ZIKV infection and the development of microcephaly was explored in a cohort of pregnant women from Rio de Janeiro, Brazil. From September 2015 to May 2016, 345 women were enrolled, of which 182 (53.0%) were positive for ZIKV in blood and/or urine samples. Development of rash 5 days before the enrollment in the study was an inclusion condition, and other common symptoms such as pruritus, arthralgia, and myalgia were also reported in most of the infected patients. The study evaluated the pregnant women who expected to deliver by the end of July 2016, and included 134 ZIKV-infected and 73 noninfected women. Between January to July 2016, among the ZIKV-infected group, of 125 available cases, 117 babies were born live from 116 pregnancies (whereas most of the miscarriages happened in the first trimester). For the noninfected group, of 61 cases, 57 babies were born. DENV infection rates were similar between the ZIKV infected and noninfected groups, while CHIKV infection rates were higher in the ZIKV noninfected than infected group. No CMV or HIV infection was detected. Microcephaly was detected in the newborns whose mothers were infected with ZIKV at week 8, 12, 30, and 38 of pregnancy, indicating no association between the period of infection and abnormal development. Newborn anomalies (mostly in the central nervous system) were detected in 42.0% (49/117) and 5% (3/57) of the infected and noninfected groups, respectively (39).

In October 2015, 31 cases of microcephaly were investigated for ZIKV infection in Pernambuco, Brazil. IgM antibodies against ZIKV were detected in CSF samples of 30 (97.0%) newborns and in serum samples of 28 (90.0%) newborns, suggesting a strong causative association of ZIKV infection with the development of microcephaly (40). Another study analyzed 1,501 newborns suspected with ZIKV infection that was reported to the Brazilian Ministry of Health from November 2015 to February 2016. From those, 899 (59.9%) cases were discarded because there was no sign of microcephaly or no additional clinical findings. The remaining 602 cases were defined as definite, high, moderate, or somewhat associated with ZIKV infection at a rate of 12.6%, 8.9%, 30.1%, or 48.4%, respectively (41). A case-control study analyzing 32 cases with microcephaly and 62 controls between January and May 2016 were performed in Brazil. None of the controls was positive for ZIKV, while 68.8% of the microcephaly cases were positive for ZIKV in CSF and/or sera. This strong association suggested that microcephaly might be caused by in utero infection, and the authors recommended adding ZIKV to the acronym associated with microcephaly, as TORCH<u>Z</u> (42).

A retrospective study conducted in Hawaii analyzed blood samples from 2009 to 2012 from pregnant women. From six mothers, whose newborns were detected with microcephaly, one (16.6%) was positive for IgM antibodies while three (50.0%) were positive for IgG antibodies against ZIKV. There was a trend indicating that a mother who was positive for ZIKV IgG or IgM antibodies was more likely to deliver a newborn with microcephaly. This study also suggested the presence of ZIKV in the United States as early as 2009 (43).

A preliminary study had analyzed images from 104 cases of microcephaly from Pernambuco, Brazil. The peak of abnormalities increased between September and October 2015. Image analysis of 58 patients revealed calcifications in most of them, mostly in cortical and subcortical junction. Among these patients, 69.0% had problems with cortical development such as pachygyria and agyria, and 66.0% had alterations in the ventricles (44).

The most common alterations in the newborn's brains, detected by computed tomography (CT) or MRI, were the presence of calcifications between cortical and subcortical white matter, which could be detected by the ultrasonogram as well. Another important finding was the absence or abnormal sulcus formation in the cortex (45, 46). The frontal part was the most affected, which was highly associated with ophthalmological problems reported in children with microcephaly (47–53). Other alterations in the corpus callosum, cisterna magna, lateral ventricles, and cerebellum were also commonly reported. Because of the microcephaly, it was common for the heads of these infants to have skin folds. How ZIKV infection affects postnatal neurological development is still unknown, but some studies have suggested that the infant might develop microcephaly after birth even if the newborn has a normal size of head circumference (45, 46).

Association between the timeline for the pregnant woman to be infected and the severity of microcephaly was investigated, although the study did not report a significant correlation (54). However, other studies reported a timeline association of ZIKV infection with microcephaly. A retrospective study was carried out to analyze ZIKV infection in the population from French Polynesia during the ZIKV outbreak from October 2013 to April 2014, and the related microcephaly cases from September 2013 to July 2015. In the 2013 outbreak, 65.0% of the population was infected, and 8 microcephaly cases were identified. Seven of them (88.0%) occurred 4 months after the end of the epidemy. By analyzing the period of ZIKV infection and the pregnancy stage, the study estimated the timeline of infection and the impact according to the gestational trimester. It was suggested that there was a 1% risk of microcephaly if the pregnant woman were infected in the first trimester. This model was designed with numerous assumptions: There was a differential risk of developing microcephaly according to the gestational stage; the microcephaly cases were all identified in that period; the number of cases detected and suspected with ZIKV were comparable for each period; the number of newborns and infected women that might get pregnant were proportional; and the birth rate did not change (9).

A study in Colombia investigated the association between the time of infection and the likelihood of developing microcephaly. From January to November 2016, 476 cases of microcephaly were reported, which were four times more than the same period in the previous year, and 147 of 306 cases (48%) were positive for ZIKV. Twenty-six out of 121 (21.0%) newborns with microcephaly were also positive for other infectious agents. An estimation between the time of infection based on the mother symptoms and the peak of ZIKV outbreak suggested a higher risk of developing microcephaly if the mother were infected in the first trimester of pregnancy (55). Other studies also associated early ZIKV infection (first and second trimester) with a higher rate of developing microcephaly (56). One important point for evaluating the timeline association is the fact that if the woman is in early pregnancy, ZIKV infection might result in miscarriage, which can be unnoticed by the mother or underreported to medical centers and hospitals. Additionally, ZIKV infection has been associated with hydrops fetalis and fetal demise (57).

The causative effect of ZIKV infection on microcephaly was extensively discussed (58–62), and many studies have showed the association between ZIKV infection and microcephaly development (8, 38, 63–66). On April 13, 2016, the US Centers for Disease Control (CDC) confirmed the link between ZIKV infection and microcephaly, congenital blindness, stillbirth, and other anomalies detected in the fetus or newborn (67). A recent report indicated that in United States, around 10.0% of the infected pregnant women who might have fetuses/infants with birth defects were associated with ZIKV infection. This study evaluated 1,297 pregnant women from January 15 to December 27, 2016 (68).

Studies focusing on ZIKV infection and the development of neural progenitor cells (hNPC) explored how the virus might modify important cellular pathways, particularly those related to differentiation, proliferation, and survival. Tang et al. (2016) showed that after infection with MR766 ZIKV strain, hNPC had an alteration in the cell cycle progression, resulting in a reduced cell growth rate. Additionally, there was an increase in the activated caspase-3, an important marker of apoptosis(69). Human-induced pluripotent stem cells (iPSC) have been used to understand how ZIKV might influence different stages of brain development in the fetus, since these cells can generate hNSC cells, neurospheres and brain organoids. Garcez et al. (2016) demonstrated MR766 ZIKV strain promoted an increase in death of hNSC with activation of caspases-3 and -7. Pyknotic nuclei were detected in the infected neurospheres. Infected brain organoids showed a 40.0% reduction in growth area when compared with the noninfected counterparts. None of these effects was observed with DENV-2 infection, indicating they were not common characteristics of infections by flaviviruses (70). ZIKV strain isolated from Brazil was also able to reduce the growth rate of the infected neurospheres (71).

Alterations in the development of brain cortical might be the key to connect ZIKV infection to microcephaly. For this reason, forebrain organoids derived from hiPSCs were infected with the MR766 ZIKV strain, which resulted in size decrease and a thinner ventricular zone-like layer. An increase in cell death and a decrease in proliferation of neural progenitor cells, and an increase in the lumen size in the ventricular structures, were also observed. Interestingly, these characteristics resembled the ones observed in newborns with microcephaly. ZIKV tropism for neural progenitor cells was also observed for hiPSCs, mimicking the first and second trimesters of human fetus development. There was an increase in the infected cells, indicating a productive infection, which resulted in the spread of the virus in the cell culture. When the Asian strain FSS13025, which was highly similar to the one isolated from Brazil, was used, there was no difference in the results observed (72).

Another study analyzing hNSC-derived from iPSC focused on the Sox2$^+$ hNSCs, an important transcription factor involved in initiation and maintenance of hNSC properties during neural differentiation. A ZIKV strain isolated from Brazil was used for infection, and there was a decrease in cellular proliferation and induction of autophagy. Figure 4.2, from a study by Souza et al. (2016), shows the consequences of ZIKV infection in hNSCs. Ultrastructural alterations can be noted at as early as 24 hours postinfection (h.p.i.). Autophagy marker, LC-3, can only be observed in ZIKV-infected cells (73).

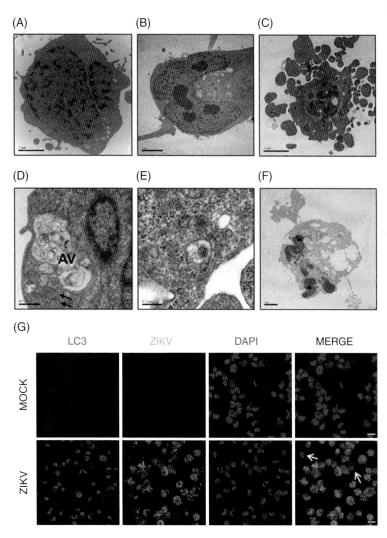

Figure 4.2 (A, B) Transmission electron micrographs of mock-infected cells, showing nuclei and organelles with normal aspect after 24 and 72 hours of culture. (C–F) ZIKV-infected cells 24 hours (C, D), 48 hours (E) and 72 (F) hours after infection. Cells in early (C) and late (F) apoptotic processes. (D) The presence of large perinuclear autophagic vacuoles (AV) can be observed. Black arrows indicate mitochondria with altered morphology. (E) Presence of viral capsid in intracellular vacuole. (G) Immunostaining for ZIKV (green) and the autophagic vacuole marker LC3 (red). Nuclei are stained with DAPI (blue). White arrows indicate cells with negative or low ZIKV staining, and absence of perinuclear LC3 staining. Scale bars = 20 μm. *Source:* Figure and legend extracted from Souza et al. (2016) (73). (*See insert for color representation of the figure.*)

Brain organoids derived from iPSCs infected with Asian strains of ZIKV isolated in French Polynesia or Brazil (H/PF/2013 or FB-GWUH-2016, respectively) caused premature differentiation of neural progenitors. Compared with the noninfected organoids, the main consequences were alterations in the centrosomes, ventricular zone, and neurogenesis, which resulted in a thinner cortical. Additionally, NSCs also diminished. Interestingly, these results were not observed when MR766 African strain was used for infection, which was consistent with the observed differential effects of different ZIKV strains in populations from Africa, South Asia and America (74).

Other studies have used hNPCs derived from iPSCs to analyze pathways that might be involved in microcephaly, including activation of p53 (75, 76) and Toll-like receptor 3 (TLR3) (77). All these data provide strong evidence to support an association of ZIKV infection with microcephaly. To further confirm these findings, many animal models have been developed. The results of these studies are discussed in the next chapter.

References

1 Oehler E, Watrin L, Larre P, Leparc-Goffart I, Lastere S, Valour F, Baudouin L, Mallet H, Musso D, Ghawche F. 2014. Zika virus infection complicated by Guillain-Barre syndrome—case report, French Polynesia, December 2013. Euro Surveill **19**.

2 Araujo LM, Ferreira ML, Nascimento OJ. 2016. Guillain-Barre syndrome associated with the Zika virus outbreak in Brazil. Arq Neuropsiquiatr **74**:253–255.

3 Malkki H. 2016. CNS infections: Zika virus infection could trigger Guillain-Barre syndrome. Nat Rev Neurol **12**:187.

4 Watrin L, Ghawche F, Larre P, Neau JP, Mathis S, Fournier E. 2016. Guillain-Barre Syndrome (42 Cases) Occurring During a Zika Virus Outbreak in French Polynesia. Medicine (Baltimore) **95**:e3257.

5 Roze B, Najioullah F, Ferge JL, Apetse K, Brouste Y, Cesaire R, Fagour C, Fagour L, Hochedez P, Jeannin S, Joux J, Mehdaoui H, Valentino R, Signate A, Cabie A, Group GBSZW. 2016. Zika virus detection in urine from patients with Guillain-Barre syndrome on Martinique, January 2016. Euro Surveill **21**.

6 Cao-Lormeau VM, Blake A, Mons S, Lastere S, Roche C, Vanhomwegen J, Dub T, Baudouin L, Teissier A, Larre P, Vial AL, Decam C, Choumet V, Halstead SK, Willison HJ, Musset L, Manuguerra JC, Despres P, Fournier E, Mallet HP, Musso D, Fontanet A, Neil J, Ghawche F. 2016. Guillain-Barre Syndrome outbreak associated with Zika virus infection in French Polynesia: a case-control study. Lancet **387**:1531–1539.

7 Ferreira MLB. 2016. Neurologic Manifestations of Arboviruses in the Epidemic in Pernambuco, Brazil. AAN 68th Annual Meeting Abstract.

8 Schuler-Faccini L, Ribeiro EM, Feitosa IM, Horovitz DD, Cavalcanti DP, Pessoa A, Doriqui MJ, Neri JI, Neto JM, Wanderley HY, Cernach M, El-Husny AS, Pone MV, Serao CL, Sanseverino MT, Brazilian Medical Genetics Society-Zika Embryopathy Task F. 2016. Possible Association Between Zika Virus Infection and Microcephaly—Brazil, 2015. MMWR Morb Mortal Wkly Rep **65**:59–62.

9 Cauchemez S, Besnard M, Bompard P, Dub T, Guillemette-Artur P, Eyrolle-Guignot D, Salje H, Van Kerkhove MD, Abadie V, Garel C, Fontanet A, Mallet H-P. Association between Zika virus and microcephaly in French Polynesia, 2013-15: a retrospective study. The Lancet **387**:2125–2132.

10 Broutet N, Krauer F, Riesen M, Khalakdina A, Almiron M, Aldighieri S, Espinal M, Low N, Dye C. 2016. Zika Virus as a Cause of Neurologic Disorders. New England Journal of Medicine **374**:1506–1509.

11 Cao-Lormeau VM, Roche C, Teissier A, Robin E, Berry AL, Mallet HP, Sall AA, Musso D. 2014. Zika virus, French polynesia, South pacific, 2013. Emerg Infect Dis **20**:1085–1086.

12 Arzuza-Ortega L, Polo A, Perez-Tatis G, Lopez-Garcia H, Parra E, Pardo-Herrera LC, Rico-Turca AM, Villamil-Gomez W, Rodriguez-Morales AJ. 2016. Fatal Sickle Cell Disease and Zika Virus Infection in Girl from Colombia. Emerg Infect Dis **22**:925–927.

13 Sarmiento-Ospina A, Vasquez-Serna H, Jimenez-Canizales CE, Villamil-Gomez WE, Rodriguez-Morales AJ. 2016. Zika virus associated deaths in Colombia. Lancet Infect Dis **16**:523–524.

14 Munoz LS, Barreras P, Pardo CA. 2016. Zika Virus-Associated Neurological Disease in the Adult: Guillain-Barre Syndrome, Encephalitis, and Myelitis. Semin Reprod Med **34**:273–279.

15 Feasby TE, Gilbert JJ, Brown WF, Bolton CF, Hahn AF, Koopman WF, Zochodne DW. 1986. An acute axonal form of Guillain-Barre polyneuropathy. Brain **109** (Pt 6)**:**1115–1126.

16 Guillain G, Barre JA, Strohl A. 1999. [Radiculoneuritis syndrome with hyperalbuminosis of cerebrospinal fluid without cellular reaction. Notes on clinical features and graphs of tendon reflexes. 1916]. Ann Med Interne (Paris) **150**:24–32.

17 Goodfellow JA, Willison HJ. 2016. Guillain-Barre syndrome: a century of progress. Nat Rev Neurol **12**:723–731.

18 Willison HJ, Jacobs BC, van Doorn PA. 2016. Guillain-Barre syndrome. Lancet **388**:717–727.

19 Cao-Lormeau V-M, Blake A, Mons S, Lastère S, Roche C, Vanhomwegen J, Dub T, Baudouin L, Teissier A, Larre P, Vial A-L, Decam C, Choumet V, Halstead SK, Willison HJ, Musset L, Manuguerra J-C, Despres P, Fournier E, Mallet H-P, Musso D, Fontanet A, Neil J, Ghawché F. Guillain-Barre Syndrome outbreak associated with Zika virus infection in French Polynesia: a case-control study. The Lancet **387**:1531–1539.

20 Brasil P, Sequeira PC, Freitas AD, Zogbi HE, Calvet GA, de Souza RV, Siqueira AM, de Mendonca MC, Nogueira RM, de Filippis AM, Solomon T. 2016. Guillain-Barre syndrome associated with Zika virus infection. Lancet **387**:1482.

21 do Rosario MS, de Jesus PA, Vasilakis N, Farias DS, Novaes MA, Rodrigues SG, Martins LC, Vasconcelos PF, Ko AI, Alcantara LC, de Siqueira IC. 2016. Guillain-Barre Syndrome After Zika Virus Infection in Brazil. Am J Trop Med Hyg **95**:1157–1160.

22 Paploski IA, Prates AP, Cardoso CW, Kikuti M, Silva MM, Waller LA, Reis MG, Kitron U, Ribeiro GS. 2016. Time Lags between Exanthematous Illness Attributed to Zika Virus, Guillain-Barre Syndrome, and Microcephaly, Salvador, Brazil. Emerg Infect Dis **22**:1438–1444.

23 Arias A, Torres-Tobar L, Hernandez G, Paipilla D, Palacios E, Torres Y, Duran J, Ugarte US, Ardila-Sierra A, Castellanos G. 2017. Guillain-Barre syndrome in patients with a recent history of Zika in Cucuta, Colombia: A descriptive case series of 19 patients from December 2015 to March 2016. J Crit Care **37**:19–23.

24 Parra B, Lizarazo J, Jimenez-Arango JA, Zea-Vera AF, Gonzalez-Manrique G, Vargas J, Angarita JA, Zuniga G, Lopez-Gonzalez R, Beltran CL, Rizcala KH, Morales MT, Pacheco O, Ospina ML, Kumar A, Cornblath DR, Munoz LS, Osorio L, Barreras P, Pardo CA. 2016. Guillain-Barre Syndrome Associated with Zika Virus Infection in Colombia. N Engl J Med **375**:1513–1523.

25 Langerak T, Yang H, Baptista M, Doornekamp L, Kerkman T, Codrington J, Roosblad J, Vreden SG, De Bruin E, Mogling R, Jacobs BC, Pas SD, GeurtsvanKessel CH, Reusken CB, Koopmans MP, Van Gorp EC, Alberga H. 2016. Zika Virus Infection and Guillain-Barre Syndrome in Three Patients from Suriname. Front Neurol 7:233.

26 Fabrizius RG, Anderson K, Hendel-Paterson B, Kaiser RM, Maalim S, Walker PF. 2016. Guillain-Barre Syndrome Associated with Zika Virus Infection in a Traveler Returning from Guyana. Am J Trop Med Hyg **95**:1161–1165.

27 Kassavetis P, Joseph JM, Francois R, Perloff MD, Berkowitz AL. 2016. Zika virus-associated Guillain-Barre syndrome variant in Haiti. Neurology **87**:336–337.

28 Dirlikov E, Major CG, Mayshack M, Medina N, Matos D, Ryff KR, Torres-Aponte J, Alkis R, Munoz-Jordan J, Colon-Sanchez C, Salinas JL, Pastula DM, Garcia M, Segarra MO, Malave G, Thomas DL, Rodriguez-Vega GM, Luciano CA, Sejvar J, Sharp TM, Rivera-Garcia B. 2016. Guillain-Barre Syndrome During Ongoing Zika Virus Transmission—Puerto Rico, January 1–July 31, 2016. MMWR Morb Mortal Wkly Rep **65**:910–914.

29 Dos Santos T, Rodriguez A, Almiron M, Sanhueza A, Ramon P, de Oliveira WK, Coelho GE, Badaro R, Cortez J, Ospina M, Pimentel R, Masis R, Hernandez F, Lara B, Montoya R, Jubithana B, Melchor A, Alvarez A, Aldighieri S, Dye C, Espinal MA. 2016. Zika Virus and the Guillain-Barre Syndrome - Case Series from Seven Countries. N Engl J Med **375**:1598–1601.

30 Leis AA, Stokic DS. 2016. Zika Virus and Guillain-Barre Syndrome: Is There Sufficient Evidence for Causality? Front Neurol 7:170.

31 Avelino-Silva VI, Martin JN. 2016. Association between Guillain-Barre syndrome and Zika virus infection. Lancet **387**:2599.

32 Sniderman A. 2010. Abnormal head growth. Pediatr Rev **31**:382–384.

33 Victora CG, Schuler-Faccini L, Matijasevich A, Ribeiro E, Pessoa A, Barros FC. 2016. Microcephaly in Brazil: how to interpret reported numbers? Lancet **387**:621–624.

34 Neu N, Duchon J, Zachariah P. 2015. TORCH infections. Clin Perinatol **42**:77–103, viii.

35 Rasmussen SA, Jamieson DJ, Honein MA, Petersen LR. 2016. Zika Virus and Birth Defects— Reviewing the Evidence for Causality. New England Journal of Medicine **374**:1981–1987.

36 Mlakar J, Korva M, Tul N, Popovic M, Poljsak-Prijatelj M, Mraz J, Kolenc M, Resman Rus K, Vesnaver Vipotnik T, Fabjan Vodusek V, Vizjak A, Pizem J, Petrovec M, Avsic Zupanc T. 2016. Zika Virus Associated with Microcephaly. N Engl J Med **374**:951–958.

37 Calvet G, Aguiar RS, Melo ASO, Sampaio SA, de Filippis I, Fabri A, Araujo ESM, de Sequeira PC, de Mendonça MCL, de Oliveira L, Tschoeke DA, Schrago CG, Thompson FL, Brasil P, dos Santos FB, Nogueira RMR, Tanuri A, de Filippis AMB. Detection and sequencing of Zika virus from amniotic fluid of fetuses with microcephaly in Brazil: a case study. The Lancet Infectious Diseases **16**:653–660.

38 Driggers RW, Ho C-Y, Korhonen EM, Kuivanen S, Jääskeläinen AJ, Smura T, Rosenberg A, Hill DA, DeBiasi RL, Vezina G, Timofeev J, Rodriguez FJ, Levanov L, Razak J, Iyengar P, Hennenfent A, Kennedy R, Lanciotti R, du Plessis A, Vapalahti O. 2016. Zika Virus Infection with Prolonged Maternal Viremia and Fetal Brain Abnormalities. New England Journal of Medicine **374**:2142–2151.

39 Brasil P, Pereira JP, Jr., Moreira ME, Ribeiro Nogueira RM, Damasceno L, Wakimoto M, Rabello RS, Valderramos SG, Halai UA, Salles TS, Zin AA, Horovitz D, Daltro P, Boechat M, Raja Gabaglia C, Carvalho de Sequeira P, Pilotto JH, Medialdea-Carrera R, Cotrim da Cunha D, Abreu de Carvalho LM, Pone M, Machado Siqueira A, Calvet GA, Rodrigues Baiao AE, Neves ES, Nassar de Carvalho PR, Hasue RH, Marschik PB, Einspieler C, Janzen C, Cherry JD, Bispo de Filippis AM, Nielsen-Saines K. 2016. Zika Virus Infection in Pregnant Women in Rio de Janeiro. N Engl J Med **375**:2321–2334.

40 Cordeiro MT, Pena LJ, Brito CA, Gil LH, Marques ET. Positive IgM for Zika virus in the cerebrospinal fluid of 30 neonates with microcephaly in Brazil. The Lancet **387**:1811–1812.

41 Franca GV, Schuler-Faccini L, Oliveira WK, Henriques CM, Carmo EH, Pedi VD, Nunes ML, Castro MC, Serruya S, Silveira MF, Barros FC, Victora CG. 2016. Congenital Zika virus syndrome in Brazil: a case series of the first 1501 livebirths with complete investigation. Lancet **388**:891–897.

42 de Araujo TV, Rodrigues LC, de Alencar Ximenes RA, de Barros Miranda-Filho D, Montarroyos UR, de Melo AP, Valongueiro S, de Albuquerque MF, Souza WV, Braga C, Filho SP, Cordeiro MT, Vazquez E, Di Cavalcanti Souza Cruz D, Henriques CM, Bezerra LC, da Silva Castanha PM, Dhalia R, Marques-Junior ET, Martelli CM, investigators from the Microcephaly Epidemic Research G, Brazilian

Ministry of H, Pan American Health O, Instituto de Medicina Integral Professor Fernando F, State Health Department of P. 2016. Association between Zika virus infection and microcephaly in Brazil, January to May, 2016: preliminary report of a case-control study. Lancet Infect Dis **16**:1356–1363.

43 Kumar M, Ching L, Astern J, Lim E, Stokes AJ, Melish M, Nerurkar VR. 2016. Prevalence of Antibodies to Zika Virus in Mothers from Hawaii Who Delivered Babies with and without Microcephaly between 2009-2012. PLoS Negl Trop Dis **10**:e0005262.

44 Microcephaly Epidemic Research G. 2016. Microcephaly in Infants, Pernambuco State, Brazil, 2015. Emerg Infect Dis **22**.

45 de Fatima Vasco Aragao M, van der Linden V, Brainer-Lima AM, Coeli RR, Rocha MA, Sobral da Silva P, Durce Costa Gomes de Carvalho M, van der Linden A, Cesario de Holanda A, Valenca MM. 2016. Clinical features and neuroimaging (CT and MRI) findings in presumed Zika virus related congenital infection and microcephaly: retrospective case series study. BMJ **353**:i1901.

46 Guillemette-Artur P, Besnard M, Eyrolle-Guignot D, Jouannic JM, Garel C. 2016. Prenatal brain MRI of fetuses with Zika virus infection. Pediatr Radiol **46**:1032–1039.

47 Miranda HA, 2nd, Costa MC, Frazao MA, Simao N, Franchischini S, Moshfeghi DM. 2016. Expanded Spectrum of Congenital Ocular Findings in Microcephaly with Presumed Zika Infection. Ophthalmology **123**:1788–1794.

48 Moshfeghi DM, de Miranda HA, 2nd, Costa MC. 2016. Zika Virus, Microcephaly, and Ocular Findings. JAMA Ophthalmol **134**:945.

49 Vasconcelos-Santos DV, Andrade GM, Caiaffa WT. 2016. Zika Virus, Microcephaly, and Ocular Findings. JAMA Ophthalmol **134**:946.

50 Ventura CV, Maia M, Dias N, Ventura LO, Belfort R, Jr. 2016. Zika: neurological and ocular findings in infant without microcephaly. Lancet **387**:2502.

51 de Oliveira Dias JR, Ventura CV, Borba PD, de Paula Freitas B, Pierroti LC, do Nascimento AP, de Moraes NS, Maia M, Belfort R, Jr. 2017. Infants with Congenital Zika Syndrome and Ocular Findings from Sao Paulo, Brazil: Spread of Infection. Retin Cases Brief Rep doi:10.1097/ICB.0000000000000518.

52 Roach T, Alcendor DJ. 2017. Zika virus infection of cellular components of the blood-retinal barriers: implications for viral associated congenital ocular disease. J Neuroinflammation **14**:43.

53 de Paula Freitas B, de Oliveira Dias JR, Prazeres J, Sacramento GA, Ko AI, Maia M, Belfort R, Jr. 2016. Ocular Findings in Infants With Microcephaly Associated With Presumed Zika Virus Congenital Infection in Salvador, Brazil. JAMA Ophthalmol doi:10.1001/jamaophthalmol.2016.0267.

54 Brasil P, Calvet GA, Siqueira AM, Wakimoto M, de Sequeira PC, Nobre A, Quintana Mde S, Mendonca MC, Lupi O, de Souza RV, Romero C, Zogbi H, Bressan Cda S, Alves SS, Lourenco-de-Oliveira R, Nogueira RM, Carvalho MS, de Filippis AM, Jaenisch T. 2016. Zika Virus Outbreak in Rio de Janeiro, Brazil: Clinical Characterization, Epidemiological and Virological Aspects. PLoS Negl Trop Dis **10**:e0004636.

55 Cuevas EL, Tong VT, Rozo N, Valencia D, Pacheco O, Gilboa SM, Mercado M, Renquist CM, Gonzalez M, Ailes EC, Duarte C, Godoshian V, Sancken CL, Turca AM, Calles DL, Ayala M, Morgan P, Perez EN, Bonilla HQ, Gomez RC, Estupinan AC, Gunturiz ML, Meaney-Delman D, Jamieson DJ, Honein MA, Martinez ML. 2016. Preliminary Report of Microcephaly Potentially Associated with Zika Virus Infection During Pregnancy—Colombia, January–November 2016. MMWR Morb Mortal Wkly Rep **65**:1409–1413.

56 Vargas A, Saad E, Dimech GS, Santos RH, Sivini MA, Albuquerque LC, Lima PM, Barreto IC, Andrade ME, Estima NM, Carvalho PI, Azevedo RS, Vasconcelos RC, Assuncao RS, Frutuoso LC, Carmo GM, Souza PB, Wada MY, Oliveira WK, Henriques CM, Percio J. 2016. Characteristics of the first cases of microcephaly possibly related to Zika virus reported in the Metropolitan Region of Recife, Pernambuco State, Brazil. Epidemiol Serv Saude **25**:691–700.

57 Sarno M, Sacramento GA, Khouri R, do Rosario MS, Costa F, Archanjo G, Santos LA, Nery N, Jr., Vasilakis N, Ko AI, de Almeida AR. 2016. Zika Virus Infection and Stillbirths: A Case of Hydrops Fetalis, Hydranencephaly and Fetal Demise. PLoS Negl Trop Dis **10**:e0004517.

58 Johansson MA, Mier-y-Teran-Romero L, Reefhuis J, Gilboa SM, Hills SL. 2016. Zika and the Risk of Microcephaly. N Engl J Med **375**:1–4.

59 Liuzzi G, Puro V, Lanini S, Vairo F, Nicastri E, Capobianchi MR, Di Caro A, Piacentini M, Zumla A, Ippolito G. 2016. Zika virus and microcephaly: is the correlation causal or coincidental? New Microbiol **39**:83–85.

60 Tetro JA. 2016. Zika and microcephaly: causation, correlation, or coincidence? Microbes Infect **18**:167–168.

61 Wang JN, Ling F. 2016. Zika Virus Infection and Microcephaly: Evidence for a Causal Link. Int J Environ Res Public Health **13**.

62 Frank C, Faber M, Stark K. 2016. Causal or not: applying the Bradford Hill aspects of evidence to the association between Zika virus and microcephaly. EMBO Molecular Medicine **8**:305–307.

63 Moron AF, Cavalheiro S, Milani H, Sarmento S, Tanuri C, de Souza FF, Richtmann R, Witkin SS. 2016. Microcephaly associated with maternal Zika virus infection. BJOG **123**:1265–1269.

64 van der Linden V, Pessoa A, Dobyns W, Barkovich AJ, Junior HV, Filho EL, Ribeiro EM, Leal MC, Coimbra PP, Aragao MF, Vercosa I, Ventura C, Ramos RC, Cruz DD, Cordeiro MT, Mota VM, Dott M, Hillard C, Moore CA. 2016. Description of 13 Infants Born During October 2015–January 2016 With Congenital Zika Virus Infection Without Microcephaly at Birth—Brazil. MMWR Morb Mortal Wkly Rep **65**:1343–1348.

65 WHO. 2016. World Health Organization. Zika situation report: Zika virus, Microcephaly and Guillain-Barré syndrome (21 April 2016). http://apps.who.int/iris/bitstream/10665/205505/1/zikasitrep_21Apr2016_eng.pdf?ua=1.

66 Hazin AN, Poretti A, Turchi Martelli CM, Huisman TA, Microcephaly Epidemic Research G, Di Cavalcanti Souza Cruz D, Tenorio M, van der Linden A, Pena LJ, Brito C, Gil LH, de Barros Miranda-Filho D, Marques ET, Alves JG. 2016. Computed Tomographic Findings in Microcephaly Associated with Zika Virus. N Engl J Med **374**:2193–2195.

67 Dyer O. 2016. US agency says Zika virus causes microcephaly. BMJ **353**:i2167.

68 Reynolds MR, Jones AM, Petersen EE, Lee EH, Rice ME, Bingham A, Ellington SR, Evert N, Reagan-Steiner S, Oduyebo T, Brown CM, Martin S, Ahmad N, Bhatnagar J, Macdonald J, Gould C, Fine AD, Polen KD, Lake-Burger H, Hillard CL, Hall N, Yazdy MM, Slaughter K, Sommer JN, Adamski A, Raycraft M, Fleck-Derderian S, Gupta J, Newsome K, Baez-Santiago M, Slavinski S, White JL, Moore CA, Shapiro-Mendoza CK, Petersen L, Boyle C, Jamieson DJ, Meaney-Delman D, Honein MA. 2017. Vital Signs: Update on Zika Virus-Associated Birth Defects and Evaluation of All U.S. Infants with Congenital Zika Virus Exposure—U.S. Zika Pregnancy Registry, 2016. MMWR Morb Mortal Wkly Rep **66**:366–373.

69 Tang H, Hammack C, Ogden Sarah C, Wen Z, Qian X, Li Y, Yao B, Shin J, Zhang F, Lee Emily M, Christian Kimberly M, Didier Ruth A, Jin P, Song H, Ming G-l. Zika Virus Infects Human Cortical Neural Progenitors and Attenuates Their Growth. Cell Stem Cell **18**:587–590.

70 Garcez PP, Loiola EC, Madeiro da Costa R, Higa LM, Trindade P, Delvecchio R, Nascimento JM, Brindeiro R, Tanuri A, Rehen SK. 2016. Zika virus impairs growth in human neurospheres and brain organoids. Science **352**:816–818.

71 Garcez PP, Nascimento JM, de Vasconcelos JM, Madeiro da Costa R, Delvecchio R, Trindade P, Loiola EC, Higa LM, Cassoli JS, Vitoria G, Sequeira PC, Sochacki J, Aguiar RS, Fuzii HT, de Filippis AM, da Silva Goncalves Vianez Junior JL, Tanuri A, Martins-de-Souza D, Rehen SK. 2017. Zika virus disrupts molecular fingerprinting of human neurospheres. Sci Rep **7**:40780.

72 Qian X, Nguyen HN, Song MM, Hadiono C, Ogden SC, Hammack C, Yao B, Hamersky GR, Jacob F, Zhong C, Yoon K-J, Jeang W, Lin L, Li Y, Thakor J, Berg DA, Zhang C, Kang E, Chickering M, Nauen D, Ho C-Y, Wen Z, Christian KM, Shi P-Y, Maher BJ, Wu H, Jin P, Tang H, Song H, Ming G-l. 2016. Brain-Region-Specific Organoids Using Minibioreactors for Modeling ZIKV Exposure. Cell **165**:1–17.

73 Souza BS, Sampaio GL, Pereira CS, Campos GS, Sardi SI, Freitas LA, Figueira CP, Paredes BD, Nonaka CK, Azevedo CM, Rocha VP, Bandeira AC, Mendez-Otero R, Dos Santos RR, Soares MB. 2016. Zika virus infection induces mitosis abnormalities and apoptotic cell death of human neural progenitor cells. Sci Rep **6**:39775.

74 Gabriel E, Ramani A, Karow U, Gottardo M, Natarajan K, Gooi LM, Goranci-Buzhala G, Krut O, Peters F, Nikolic M, Kuivanen S, Korhonen E, Smura T, Vapalahti O, Papantonis A, Schmidt-Chanasit J, Riparbelli M, Callaini G, Kronke M, Utermohlen O, Gopalakrishnan J. 2017. Recent Zika Virus Isolates Induce Premature Differentiation of Neural Progenitors in Human Brain Organoids. Cell Stem Cell **20**:397–406 e395.

75 Ghouzzi VE, Bianchi FT, Molineris I, Mounce BC, Berto GE, Rak M, Lebon S, Aubry L, Tocco C, Gai M, Chiotto AM, Sgro F, Pallavicini G, Simon-Loriere E, Passemard S, Vignuzzi M, Gressens P, Di Cunto F. 2016. ZIKA virus elicits P53 activation and genotoxic stress in human neural progenitors similar to mutations involved in severe forms of genetic microcephaly and p53. Cell Death Dis **7**:e2440.

76 Ghouzzi VE, Bianchi FT, Molineris I, Mounce BC, Berto GE, Rak M, Lebon S, Aubry L, Tocco C, Gai M, Chiotto AM, Sgro F, Pallavicini G, Simon-Loriere E, Passemard S, Vignuzzi M, Gressens P, Di Cunto F. 2017. ZIKA virus elicits P53 activation and genotoxic stress in human neural progenitors similar to mutations involved in severe forms of genetic microcephaly and p53. Cell Death Dis **8**:e2567.

77 Dang J, Tiwari SK, Lichinchi G, Qin Y, Patil VS, Eroshkin AM, Rana TM. 2016. Zika Virus Depletes Neural Progenitors in Human Cerebral Organoids through Activation of the Innate Immune Receptor TLR3. Cell Stem Cell **19**:258–265.

5

ZIKV Animal Models

5.1 Animal Models: Embryonated Hen Eggs

The first animal experiments developed for ZIKV dated almost 60 years ago. The study using embryonated hen eggs demonstrated that ZIKV was able to infect chick embryos. Inoculation into yolk, amniotic sacs and allantoic sacs led to persistent infection for up to 9 days but infection through allantoic sacs progressed at a slower rate since the virus was detected at 4 days post–infection (d.p.i.) compared to 2 d.p.i. by other routes. However, there was no difference in virus titer between the body and brain in the embryos inoculated by the previously mentioned routes, and the infection was not fatal (1). Curiously, 64 years after this first study, another one with embryonic eggs was published using a ZIKV strain isolated in Mexico. Lethality was observed using high doses of virus. ZIKV was able to replicate in many organs besides brain. The alterations caused in the central nervous system were similar to the microcephaly recently observed in human newborns (2).

5.2 Animal Models: Landrace Piglet

ZIKV PRVABC59 strain was able to infect 1-day old piglet. The virus was injected intracerebrally (I.C.), intradermally (I.D.), or intraperitoneally (I.P.). Neurological disorders were noted at 5 d.p.i. in 2 out of 11 piglets from the I.C. group, which was also the only group that ZIKV

Zika Virus and Diseases: From Molecular Biology to Epidemiology, First Edition.
Suzane Ramos da Silva, Fan Cheng and Shou-Jiang Gao.
© 2018 John Wiley & Sons, Inc. Published 2018 by John Wiley & Sons, Inc.

was detected in the brain. I.D. and I.P groups had ZIKV detected in serum at 3, 5, and 7 d.p.i., and at 3 and 5 d.p.i., respectively. Interestingly, all groups had ZIKV detected in the spleen (3).

5.3 Animal Models: Mice

The effect of ZIKV infection in infant mice was described by Weinbren & Williams (1958) with strains isolated from Lunyo forest. Besides death following the infection, other findings included encephalitis, skeletal myositis, and myocarditis. No lesion could be detected in other organs except edema in the lungs of the mice with myocarditis. Additionally, the virus adaption to the mice was observed since the symptoms manifested earlier after a few passages (4). Bell et al. (1971) demonstrated that ZIKV infection affected mainly the nervous central system of mice. ZIKV strain MP1751, which was isolated from *Aedes africanus*, was inoculated by I.C. into 1-day-old Webster Swiss (WS) white mice. At 7 d.p.i., the mice were sick, and their brains were collected and injected into 1-day and 5-week-old WS white mice. ZIKV could replicate in neurons and astroglial cells in these mice, and the most noticeable pathology was damage in the pyriform cell of the Ammon's horn. All newborns were sick at 6 d.p.i. while 20% of the 5-week-old were sick between 14 and 20 d.p.i. At the later time points, infected astroglial cells could be found all over the cortex, and extended beyond the Ammon's horn (5).

The first report to show that adult A129 mice lacking interferon (IFN)-α/β responses were susceptible to ZIKV infection used the Africa strain MP1751 isolated in 1960s (6). The virus was injected subcutaneously to mimic a mosquito bite. A129 mice showed drastic weight loss, which was a sign of infection, and the study endpoint was at 6 d.p.i. Virus RNA was detected in most organs at 3 d.p.i. with the highest concentration in the spleen at 6 d.p.i. The brain was the only organ that did not present any decrease in the viral load at later time points compared to 3 d.p.i. An immunocompetent parent strain of A129 mice, 129Sv/Ev, showed no sign of infection but had ZIKV RNA detection in some organs with lower viral loads than those in A129 mice. Morphological alterations of the brain were observed in A129 mice with infiltration of inflammatory cells and diffused alterations in gray and white matter (7).

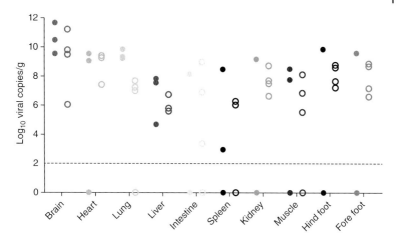

Figure 5.1 ZIKV RNA detection in different tissues extracted from mice infected with ZIKV, including young (closed symbols) and adult mice (open symbols). *Source:* Figure and legend extracted from Aliota et al. (2016) (8).

AG129 mice has also been used for ZIKV infection because of its deficiency in IFN-α/β and -γ receptors. The vulnerability of this model was described by Aliota (2016) with a ZIKV strain from French Polynesia. The virus was detected in many organs (Figure 5.1) but severe alterations were only detected in the brain. Doses varying from 1 to 1×10^5 plaque-forming units (PFU) were fatal in less than 8 d.p.i. in 3- to 4-week-old or 8-week-old mice by either I.P. or foot pad inoculation. The viral load peaked at 2 d.p.i. in both mice ages, and the weight loss reached up to 20% with other signs of illness such as lethargy and curved posture at day 5 d.p.i. even in animals that received a lower dose of ZIKV (8, 9).

BALB/c mice immunosuppressed with dexamethasone were shown to be a good model for ZIKV infection. Disseminated infection was observed in 6- to 8-week-old male and female mice. Male mice developed infection symptoms earlier than the female mice, and their clinical symptoms were aggravated with significant weight loss (more than 10%), and orchitis after dexamethasone treatment was stopped at 9 d.p.i. The survival rate of the male group was drastic improved with IFN-I treatment (10).

To understand the biology of the recently isolated Asia/American ZIKV strain, Zhang et al. (2016) analyzed the strain GZ01 isolated from Venezuela and used it to inoculate 4-week-old male 129Sv/Ev mice by I.P. Viremia peaked at 1 d.p.i. and became undetectable by 3 d.p.i. ZIKV RNA was detected in all organs at 1 d.p.i., with the highest viral load found in testis, which, together with the spleen, was still positive at 3 d.p.i. The only organ that showed alterations based on hematoxylin-eosin analysis was the liver. Similar results were observed in Babl/C mice. For 8-week-old 129Sv/Ev mice, the results were similar, but the viral peak was at 2 d.p.i (11).

Three animal models simultaneously demonstrated that abnormal brain development was caused by ZIKV infection. One study used a strain isolated from Brazil (ZIKV^{BR}) to intravenously infect two different strains of pregnant mice, SJL and C57BL/6, at around day 10 to 13 of gestation. Pups from C57BL/6 mice with normal IFN responses were not infected by ZIKV^{BR}, indicating that the virus did not cross the placenta barrier, which was consistent with previous study (12). In contrast, pups from SJL mice had intrauterine growth restriction (IUGR), which resulted in malformations. Genes involved in autophagy and apoptosis were deregulated. Additionally, neural progenitor cells (NPCs) were infected by ZIKV, and infectious viral particles were observed with 1 multiplicity of infection (MOI) while 10 MOI was necessary for infection in neurons. An increased number of cell deaths was observed in the infected NPC at 10 MOI (13). 3D cell cultures showed that at 10 MOI, neurospheres infected with ZIKV^{BR} were smaller than those infected with ZIKV^{766}. Human cerebral organoids infected with both ZIKV strains at 0.1 MOI showed a decrease in CTIP2$^+$ and PAX-6$^+$ cells, while only ZIKV^{BR} promoted a decrease of TBR1$^+$, KI67$^+$, and SOX2$^+$ cells. On the other hand, ZIKV^{BR} was not able to replicate in cerebral organoids derived from chimpanzees and did not reduce CTIP2$^+$ and TBR1$^+$ cells (13).

The second study evaluated the effect of ZIKV in mouse fetuses. An Asian ZIKV strain was injected into the cerebroventricular space/lateral ventricle of the brain at embryonic day 13.5. At 3 to 5 d.p.i., most of the ZIKV-infected cells were found in the ventricular and subventricular zones; a thinner cortical layer was detected; and smaller brains in the infected animals than those of uninfected littermates were observed. Additionally, cleaved-caspase-3 was detected in the intermediate zone and cortical plate. ZIKV preferentially infected NPCs and intermediate or basal progenitor cells (iPSCs/bPSCs), and inhibited the proliferation and differentiation of NPCs. RNA-seq analysis identified an upregulation of pathways associated with immune response and apoptosis and downregulation of those related to cell proliferation and differentiation (14).

A third study evaluated the effect of S.C. infection with the Asian ZIKV strain (H/PF/2013) at gestation day 6.5 and 7.5 of heterozygous fetuses ($Ifnar1^{+/-}$) derived from backcross of $Ifnar1^{-/-}$ and C57BL/6 mice. Infection of these fetuses, which had type I IFN response, resulted in fetal death and reabsorption in most of them, while the ones that survived infection had IUGR and growth impairment. ZIKV RNA detected was significantly higher in the placenta than in maternal sera. While mice treated with a blocking antibody against IFN-α receptor prior exposure to ZIKV infection developed IUGR, there was no fetal demise. Important vascular alterations were detected in the placenta, and ZIKV was mainly detected in glucogen trophoblasts and spongiotrophoblast cells (15).

ZIKV infection at some specific brain stages could enhance the likelihood of microcephaly. Wild type mice C57BL/6 of 1- and 3 weeks old, representing brain development at early and adult brain stages, respectively, were injected via intracranial route. Although it was extensively found in brains in both groups of animals, the younger mice showed more apoptosis in specific cell types such as corticospinal pyramidal neurons than the older ones did, besides the decrease in proliferating cells in the ventricular zone of the brains (16).

The ability of ZIKV to infect the brain neocortex was demonstrated by Brault et al. (2016) using E15 mouse embryonic brain and French Polynesia ZIKV strain (H/PF/2013). The virus mostly infected radial glial progenitor (RGP) cells, caused a block in cellular proliferation without inducing apoptosis at the early infection time such as 24 h.p.i. (17). A ZIKV strain isolated from Brazil (SPH 2015) was also used for infection in Swiss newborn mice by I.C. or S.C. in the lumbar area (18). Both groups presented damage in CNS. The cortex was the most affected area, and the hippocampus had moderate/diffuse or discrete/segmental necrosis. Myelopathy was only observed in the group infected by I.C., while astrogliosis with overexpression of glial fibrillary acidic protein was detected in both groups (18). ZIKV infection was also investigated in neonatal C57BL/6 mice and C57BL/6 mice with IFNAR knockout by subcutaneous inoculation using ZIKV strain PRVABC59. The infection was fatal for the knockout mice at 5 to 6 d.p.i., while the counterpart wild-type mice presented neurological alterations. Interestingly, C57BL/6 mice had mostly CD8+ T cells infiltrations while the knockout mice had inflammatory cells in the CNS (19). Alterations in the formation of eyes caused by ZIKV were observed in 4- to 6-week-old wild-type C57BL/6 and mice with IFN-stimulated gene 15 (ISG15) knockout intravitreally infected with PRVABC59. ZIKV was able to infect retinal endothelium and pigment epithelium, resulting in retinal damage as a result of cell death. The ISG15 knockout mice showed more serious chorioretinitis than the wild-type mice did, indicating a prominent role of ISG15 in controlling ZIKV infection (20).

Other IFN-α/β knockout mice (Irf3$^{-/-}$Irf5$^{-/-}$Irf7$^{-/-}$ and Ifnar1$^{-/-}$) were also susceptible to African (MR766, Dakar41671, Dakar 41667, and Dakar 41519) and Asian (H/PF/2013) ZIKV strains. All mice lacking IFN-α/β response had higher ZIKV viral loads than adult wild-type mice (C57BL/6). However, 1-week-old C57BL/6 mice were quite susceptible to ZIKV infection. Interestingly, morbidity and mortality rates were higher in mice infected by H/PF/2013 than MR766 ZIKV strain. High viral loads were detected at 6 d.p.i. in brain, spinal cord, and testes of 4- to 6-week-old Ifnar1$^{-/-}$ mice infected S.C. with H/PF/2013. ZIKV RNA could be detected for up to 28 d.p.i. in brain and testes (21). Another study using 6-week-old Irf3$^{-/-}$Irf5$^{-/-}$Irf7$^{-/-}$ triple knockout mice, infected via retro-orbital with Asian ZIKV strain (FSS13025), demonstrated ZIKV infection of neural progenitor cells in the subventricular and subgranular zones of the brain, resulting in apoptosis and impairment of cell proliferation (22).

Teratogenic effects of ZIKV infection were analyzed in immuno-competent mice FVB/NJ and C57BL/6J. The mice were infected via jugular venous with ZIKV HS-2015-BA-01 isolated in Brazil. Pregnant females presented ZIKV RNA mostly in spleens and kidneys, while all blood samples were negative. Mice in the embryos presented dys-raphia and hydrocephalus at 5 days of developmental stage and serious malformations while at 10.5 days post-coitus (d.p.c). ZIKV infection of mice between 7.5-9.5 d.p.c. resulted in intrauterine growth restriction (IUGR) and malformations while no development problem was observed for those infected at 12.5 d.p.c. (Figure 5.2) (23).

5.4 Animal Model: Nonhuman Primate

Animal studies have provided relevant information about the patho-genesis of ZIKV infection. One important difference that must be considered is the gestational period between the animal used and

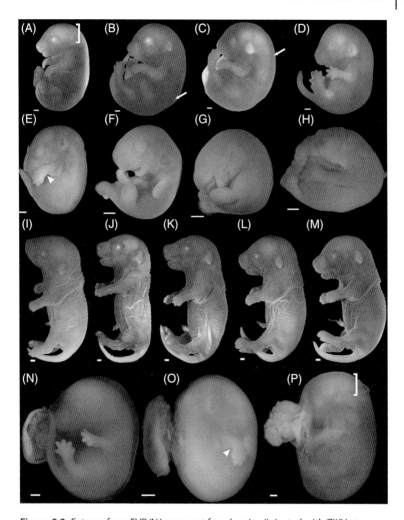

Figure 5.2 Fetuses from FVB/NJ pregnant female mice (injected with ZIKV at 9.5 days post-coitus (d.p.c.), and harvested at 16.5 d.p.c.) organized in order of slight to severe malformation (A-H). Littermates with normal (I-M) and abnormal (N-P) phenotypes. The white arrowhead indicates an atypical forelimb posture. Scale bar, 2.0 mm. *Source:* Figure and legends extracted from Xavier-Neto et al. (2016) (23).

human, especially between mice and humans (~21 *vs* ~280 days), when analyzing ZIKV infection in different stages of pregnancy. The pregnancy time will influence the period needed for brain development, and consequently, it is complicated to compare the impact of ZIKV infection among different species. When comparing all the animal models, two important points must be considered: strain and dose of virus used in the experiments. For all the listed reasons, the model based on nonhuman primates is extremely important and relevant.

Dudley et al. (2016) have shown that rhesus macaque can be subcutaneously infected by Asian ZIKV strain (H/PF/2013). Six male monkeys received different doses of the virus (1×10^6, 1×10^5 or 1×10^4 P.F.U.), and 2 pregnant animals received 1×10^4 P.F.U. of the virus at gestation day 31 or 38, respectively. Symptoms observed included slight weight loss because of lack of eating and a rash at the virus injection site lasting for a few days. ZIKV RNA was detected in all animals from 1 d.p.i. to 21 d.p.i. in the male animals, and up to 57 d.p.i. in the pregnant ones. The RNA was detected in saliva, urine, and cerebrospinal fluid samples. All animals showed an increase in natural killers (NKs), CD8$^+$ and CD4$^+$ T cells, and plasmablasts. All animals had neutralizing antibodies against ZIKV by 21 d.p.i. A rechallenge of three male monkeys at 10 weeks p.i. showed no virus replication (24).

A rechallenge model demonstrated protection when virus used for infection and rechallenge were different strains. Three monkeys were subcutaneously infected with African strain (MR766) at 1×10^4, 1×10^5, or 1×10^6 PFU. The re-challenge with 1×10^4 PFU of Asian ZIKV strain (H/PF/2013) was done at 70 d.p.i., and ZIKV RNA was not detected in plasma, saliva, or urine for up to 21 days after re-challenge (25).

Another study investigating ZIKV infection in 5-year-old rhesus monkeys, which were subcutaneously injected with 1×10^5 P.F.U. of a recently isolated strain from Venezuela GZ01/2016 (26). Fever was the main symptom of the infected animals, and ZIKV RNA was detected in blood, urine, and body fluids. Animals were euthanized at 5 and 10 d.p.i. ZIKV was mostly detected in the intestine, spleen, parotid glands, and lymph nodes (Figure 5.3) (26).

ZIKV pathobiology was also analyzed in a Mauritian cynomolgus macaque (MCM) model. Male and female animals were infected with 5×10^5 P.F.U. and 1×10^4 P.F.U. of ZIKV, respectively, of African (IbH30656) and Asian (PRVABC59 and FSS13025) strains. No clinical symptom was presented. Interestingly, IbH30656 strain was not able to establish infection and failed to produce neutralization antibodies. On the other hand, PRVABC59 and FSS13025 strains could induce

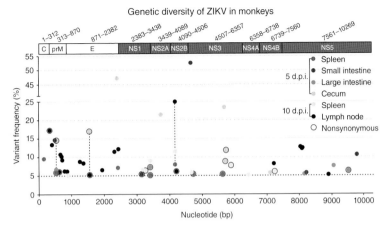

Figure 5.3 Analysis of ZIKV diversity in different tissues from monkey rhesus. Dotted lines indicate consensus changes. *Source:* Figure and legends are adapted from Li et al. (2016)(26). (*See insert for color representation of the figure.*)

viremia in sera for up to 10 d.p.i. and 6 d.p.i., respectively. The highest viral load detected was in testes from animals infected with FSS13025, while the levels were moderated in urine and saliva in both Asian groups. No virus was detected in brain samples (27).

The first nonhuman primate study to demonstrate brain fetal damage after infection with ZIKV was published by Waldorf et al. (2016). Pregnant pigtail macaques at 119 days of gestational period were subcutaneously infected with FSS13025 ZIKV strain at 1×10^7 P.F.U. for five times. It is important to note that this was the highest viral dose used in monkey models up to this point. No clinical symptom developed after the infection, and a Cesarean was performed on the animals at 162 days of gestation. The virus persisted in the mother and fetus organs for up to 6 weeks p.i., whereas the highest viral load was detected in the fetal brains. Pathological changes in the fetal brains included white matter hypoplasia and gliosis, and apoptosis in subependymal cells. Results of this study support the hypothesis that there is a causal effect between ZIKV infection in pregnant women and the development of microcephaly in the newborns (28).

A more recent study challenged four pregnant rhesus monkeys with an Asian strain (H.sapiens-tc/FRA/2013/FrenchPolynesia-01_v1c1) by S.C. injection through the cranial dorsum with 1.26×10^6 PFU in different gestational periods. Although no fetal growth restriction was observed, ZIKV RNA was detected in tissues from all the fetuses, indicating a high transmission rate (29). Viral persistence was also recently reported in rhesus monkeys infected S.C. with Asian strain (PRVABC59). The animals received up to 10 injections over hands and arms with doses of 1×10^4, 1×10^5, or 1×10^6 focus forming units (FFU). ZIKV RNA was detected for up to 35 d.p.i. in neuronal, lymphoid, and joint and muscle tissues, as well as reproductive organs (30, 31).

References

1 Taylor RM. 1952. Studies on certain viruses isolated in the tropics of Africa and South America; their growth and behavior in the embryonated hen egg. J Immunol **68**:473–494.

2 Goodfellow FT, Tesla B, Simchick G, Zhao Q, Hodge T, Brindley MA, Stice SL. 2016. Zika Virus Induced Mortality and Microcephaly in Chicken Embryos. Stem Cells Dev **25**:1691–1697.

3 Darbellay J, Lai K, Babiuk S, Berhane Y, Ambagala A, Wheler C, Wilson D, Walker S, Potter A, Gilmour M, Safronetz D, Gerdts V, Karniychuk U. 2017. Neonatal pigs are susceptible to experimental Zika virus infection. Emerg Microbes Infect **6**:e6.

4 Weinbren MP, Williams MC. 1958. Zika virus: further isolations in the Zika area, and some studies on the strains isolated. Trans R Soc Trop Med Hyg **52**:263–268.

5 Bell TM, Field EJ, Narang HK. 1971. Zika virus infection of the central nervous system of mice. Arch Gesamte Virusforsch **35**:183–193.

6 Haddow AJ, Williams MC, Woodall JP, Simpson DI, Goma LK. 1964. Twelve Isolations of Zika Virus from Aedes (Stegomyia) Africanus (Theobald) Taken in and above a Uganda Forest. Bull World Health Organ **31**:57–69.

7 Dowall SD, Graham VA, Rayner E, Atkinson B, Hall G, Watson RJ, Bosworth A, Bonney LC, Kitchen S, Hewson R. 2016. A Susceptible Mouse Model for Zika Virus Infection. PLoS Negl Trop Dis **10**:e0004658.

8 Aliota MT, Caine EA, Walker EC, Larkin KE, Camacho E, Osorio JE. 2016. Characterization of Lethal Zika Virus Infection in AG129 Mice. PLoS Negl Trop Dis **10**:e0004682.

9 Aliota MT, Caine EA, Walker EC, Larkin KE, Camacho E, Osorio JE. 2016. Correction: Characterization of Lethal Zika Virus Infection in AG129 Mice. PLoS Negl Trop Dis **10**:e0004750.

10 Chan JF, Zhang AJ, Chan CC, Yip CC, Mak WW, Zhu H, Poon VK, Tee KM, Zhu Z, Cai JP, Tsang JO, Chik KK, Yin F, Chan KH, Kok KH, Jin DY, Au-Yeung RK, Yuen KY. 2016. Zika Virus Infection in Dexamethasone-immunosuppressed Mice Demonstrating Disseminated Infection with Multi-organ Involvement Including Orchitis Effectively Treated by Recombinant Type I Interferons. EBioMedicine **14**:112–122.

11 Zhang NN, Tian M, Deng YQ, Hao JN, Wang HJ, Huang XY, Li XF, Wang YG, Zhao LZ, Zhang FC, Qin CF. 2016. Characterization of the contemporary Zika virus in immunocompetent mice. Hum Vaccin Immunother **12**:3107–3109.

12 Rossi SL, Tesh RB, Azar SR, Muruato AE, Hanley KA, Auguste AJ, Langsjoen RM, Paessler S, Vasilakis N, Weaver SC. 2016. Characterization of a Novel Murine Model to Study Zika Virus. The American Journal of Tropical Medicine and Hygiene **94**:1362–1369.

13 Cugola FR, Fernandes IR, Russo FB, Freitas BC, Dias JL, Guimaraes KP, Benazzato C, Almeida N, Pignatari GC, Romero S, Polonio CM, Cunha I, Freitas CL, Brandao WN, Rossato C, Andrade DG, Faria Dde P, Garcez AT, Buchpigel CA, Braconi CT, Mendes E, Sall AA, Zanotto PM, Peron JP, Muotri AR, Beltrao-Braga PC. 2016. The Brazilian Zika virus strain causes birth defects in experimental models. Nature **534**:267–271.

14 Li C, Xu D, Ye Q, Hong S, Jiang Y, Liu X, Zhang N, Shi L, Qin C-F, Xu Z. Zika Virus Disrupts Neural Progenitor Development and Leads to Microcephaly in Mice. Cell Stem Cell **19**:120–126.

15 Miner Jonathan J, Cao B, Govero J, Smith Amber M, Fernandez E, Cabrera Omar H, Garber C, Noll M, Klein Robyn S, Noguchi Kevin K, Mysorekar Indira U, Diamond Michael S. Zika Virus Infection during Pregnancy in Mice Causes Placental Damage and Fetal Demise. Cell **165**:1081–1091.

16 Huang WC, Abraham R, Shim BS, Choe H, Page DT. 2016. Zika virus infection during the period of maximal brain growth causes microcephaly and corticospinal neuron apoptosis in wild type mice. Sci Rep **6**:34793.

17 Brault JB, Khou C, Basset J, Coquand L, Fraisier V, Frenkiel MP, Goud B, Manuguerra JC, Pardigon N, Baffet AD. 2016. Comparative Analysis Between Flaviviruses Reveals Specific Neural Stem Cell Tropism for Zika Virus in the Mouse Developing Neocortex. EBioMedicine **10**:71–76.

18 Fernandes NC, Nogueira JS, Ressio RA, Cirqueira CS, Kimura LM, Fernandes KR, Cunha MS, Souza RP, Guerra JM. 2017. Experimental Zika virus infection induces spinal cord injury and encephalitis in newborn Swiss mice. Exp Toxicol Pathol **69**:63–71.

19 Manangeeswaran M, Ireland DD, Verthelyi D. 2016. Zika (PRVABC59) Infection Is Associated with T cell Infiltration and Neurodegeneration in CNS of Immunocompetent Neonatal C57Bl/6 Mice. PLoS Pathog **12**:e1006004.

20 Singh PK, Guest JM, Kanwar M, Boss J, Gao N, Juzych MS, Abrams GW, Yu FS, Kumar A. 2017. Zika virus infects cells lining the blood-retinal barrier and causes chorioretinal atrophy in mouse eyes. JCI Insight **2**:e92340.

21 Lazear Helen M, Govero J, Smith Amber M, Platt Derek J, Fernandez E, Miner Jonathan J, Diamond Michael S. 2016. A Mouse Model of Zika Virus Pathogenesis. Cell Host & Microbe **19**:720–730.

22 Li H, Saucedo-Cuevas L, Regla-Nava JA, Chai G, Sheets N, Tang W, Terskikh AV, Shresta S, Gleeson JG. 2016. Zika Virus Infects Neural Progenitors in the Adult Mouse Brain and Alters Proliferation. Cell Stem Cell **19**:593–598.

23 Xavier-Neto J, Carvalho M, Pascoalino BD, Cardoso AC, Costa AM, Pereira AH, Santos LN, Saito A, Marques RE, Smetana JH, Consonni SR, Bandeira C, Costa VV, Bajgelman MC, Oliveira PS, Cordeiro MT, Gonzales Gil LH, Pauletti BA, Granato DC, Paes Leme AF, Freitas-Junior L, Holanda de Freitas CB, Teixeira MM, Bevilacqua E, Franchini K. 2017. Hydrocephalus and arthrogryposis in an immunocompetent mouse model of ZIKA teratogeny: A developmental study. PLoS Negl Trop Dis **11**:e0005363.

24 Dudley DM, Aliota MT, Mohr EL, Weiler AM, Lehrer-Brey G, Weisgrau KL, Mohns MS, Breitbach ME, Rasheed MN, Newman CM, Gellerup DD, Moncla LH, Post J, Schultz-Darken N, Schotzko ML, Hayes JM, Eudailey JA, Moody MA, Permar SR, O'Connor SL, Rakasz EG, Simmons HA, Capuano S, Golos TG, Osorio JE, Friedrich TC, O'Connor DH. 2016. A rhesus macaque model of Asian-lineage Zika virus infection. Nat Commun **7**:12204.

25 Aliota MT, Dudley DM, Newman CM, Mohr EL, Gellerup DD, Breitbach ME, Buechler CR, Rasheed MN, Mohns MS, Weiler AM, Barry GL, Weisgrau KL, Eudailey JA, Rakasz EG, Vosler LJ, Post J, Capuano S, 3rd, Golos TG, Permar SR, Osorio JE, Friedrich TC, O'Connor SL, O'Connor DH. 2016. Heterologous Protection against Asian Zika Virus Challenge in Rhesus Macaques. PLoS Negl Trop Dis **10**:e0005168.

26 Li XF, Dong HL, Huang XY, Qiu YF, Wang HJ, Deng YQ, Zhang NN, Ye Q, Zhao H, Liu ZY, Fan H, An XP, Sun SH, Gao B, Fa YZ, Tong YG, Zhang FC, Gao GF, Cao WC, Shi PY, Qin CF. 2016. Characterization of a 2016 Clinical Isolate of Zika Virus in Non-human Primates. EBioMedicine **12**:170–177.

27 Koide F, Goebel S, Snyder B, Walters KB, Gast A, Hagelin K, Kalkeri R, Rayner J. 2016. Development of a Zika Virus Infection Model in Cynomolgus Macaques. Front Microbiol 7:2028.

28 Adams Waldorf KM, Stencel-Baerenwald JE, Kapur RP, Studholme C, Boldenow E, Vornhagen J, Baldessari A, Dighe MK, Thiel J, Merillat S, Armistead B, Tisoncik-Go J, Green RR, Davis MA, Dewey EC, Fairgrieve MR, Gatenby JC, Richards T, Garden GA, Diamond MS, Juul SE, Grant RF, Kuller L, Shaw DW, Ogle J, Gough GM, Lee W, English C, Hevner RF, Dobyns WB, Gale M, Jr., Rajagopal L. 2016. Fetal brain lesions after subcutaneous inoculation of Zika virus in a pregnant nonhuman primate. Nat Med **22**:1256–1259.

29 Nguyen SM, Antony KM, Dudley DM, Kohn S, Simmons HA, Wolfe B, Salamat MS, Teixeira LBC, Wiepz GJ, Thoong TH, Aliota MT, Weiler AM, Barry GL, Weisgrau KL, Vosler LJ, Mohns MS, Breitbach ME, Stewart LM, Rasheed MN, Newman CM, Graham ME, Wieben OE, Turski PA, Johnson KM, Post J, Hayes JM, Schultz-Darken N, Schotzko ML, Eudailey JA, Permar SR, Rakasz EG, Mohr EL, Capuano S, 3rd, Tarantal AF, Osorio JE, O'Connor SL, Friedrich TC, O'Connor DH, Golos TG. 2017. Highly efficient maternal-fetal Zika virus transmission in pregnant rhesus macaques. PLoS Pathog **13**:e1006378.

30 Hirsch AJ, Smith JL, Haese NN, Broeckel RM, Parkins CJ, Kreklywich C, DeFilippis VR, Denton M, Smith PP, Messer WB, Colgin LM, Ducore RM, Grigsby PL, Hennebold JD, Swanson T, Legasse AW, Axthelm MK, MacAllister R, Wiley CA, Nelson JA, Streblow DN. 2017. Correction: Zika Virus infection of rhesus macaques leads to viral persistence in multiple tissues. PLoS Pathog **13**:e1006317.

31 Hirsch AJ, Smith JL, Haese NN, Broeckel RM, Parkins CJ, Kreklywich C, DeFilippis VR, Denton M, Smith PP, Messer WB, Colgin LM, Ducore RM, Grigsby PL, Hennebold JD, Swanson T, Legasse AW, Axthelm MK, MacAllister R, Wiley CA, Nelson JA, Streblow DN. 2017. Zika Virus infection of rhesus macaques leads to viral persistence in multiple tissues. PLoS Pathog **13**:e1006219.

6

Biology of ZIKV

ZIKV is a member of the family of *Flaviviridae* (from the Latin *flavus*, "yellow"), which gets its name from the Yellow Fever virus (YFV). *Flaviviridae* family currently consists of three genera: *Flavivirus*, *Pestivirus* (from the Latin *pestis*, "plague"), and *Hepacivirus* (from the Greek *hepar*, *hepatos*, "liver"). Members of the *Flaviviridae* share similar genome organization, virion structure, and replication strategy but display diverse biological properties and lack serologic cross-reactivity (1).

ZIKV belongs to the *Flavivirus* genus, and is thus closely related to the dengue virus (DENV), YFV, Japanese encephalitis virus (JEV), and West Nile virus (WNV). Most flaviviruses are transmitted by insect vectors, which is a unique feature of the genus not shared by the pestiviruses and hepaciviruses of the Flaviviridae family (1). Based on their mode of transmission, three groups of flaviviruses are defined: tick-borne flaviviruses (TBFVs), mosquito-borne flaviviruses (MBFVs), and those flaviviruses with no known vector (NKV) (Figure 6.1) (1, 2). As a mosquito-borne virus, ZIKV is genetically and antigenically related to Spondweni virus (SPOV). They form clade X within the mosquito-borne flavivirus cluster (Figure 6.1) (1–3).

In the previous chapters, the discovery, transmission, and evolution of ZIKV as well as its associated pathogenesis have been extensively discussed. This chapter will introduce the basic biology of ZIKV, including its structural and genomic characteristics, replication cycle, and viral protein functions.

Zika Virus and Diseases: From Molecular Biology to Epidemiology, First Edition.
Suzane Ramos da Silva, Fan Cheng and Shou-Jiang Gao.
© 2018 John Wiley & Sons, Inc. Published 2018 by John Wiley & Sons, Inc.

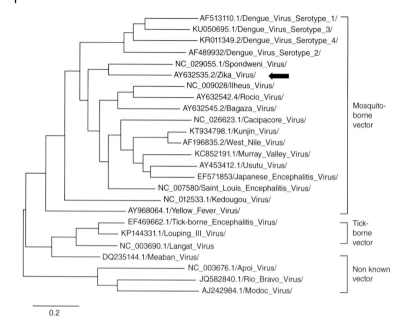

Figure 6.1 Representative phylogram of the genus Flavivirus. The phylogenetic tree was plotted based on complete NS5 nucleotide sequence built from a multiple alignment using Clustal omega and Phylogeny.fr. (235). The scale represents 0.2 substitutions per site. Accession numbers of viral genome sequences are displayed. The arrow highlights ZIKV. *Source:* Extracted from Saiz et al. (2016) (2).

6.1 Structural and Physical Properties of ZIKV Virion

Similar to other flaviviruses, ZIKV particles are enveloped and contain the envelope (E) glycoprotein, the membrane (M) protein, and the capsid (C) protein, as well as the genomic RNA. In virions, the RNA genome is complexed with multiple capsid proteins, surrounded by a membrane (4). The extracellular, mature form of the virus has an outer surface that forms an icosahedral shell with 90 dimers of E and M proteins, arranged in a herringbone pattern (5). The E protein is the major protein involved in receptor binding and fusion while the M protein is a small protein hidden under the E protein layer. They are anchored in the lipid bilayer membrane via their transmembrane regions (6).

The early transmission electron microscopy (TEM) study indicated that ZIKV virions were spherical particles with an overall diameter of 40–43 nm, and a central core of 28–30 nm in diameter (7). The cryo-EM structures of various flaviviruses, including ZIKV, have been resolved and they have a similar overall architecture (Figure 6.2) (5, 6, 8–10). When comparing the structure of ZIKV with other flaviviruses, a noticeable feature is the protruding density on the surface of the virus (red in Figure 6.2 B and C) (10), which is the glycan on the E protein. The E proteins exhibited the characteristic "herringbone" structure in the virion, where there is one dimeric heterodimer $(E-M)_2$ located on each of 30 twofold vertices and 60 dimeric heterodimers $(E-M)_2$ in general positions within the icosahedral protein shell (10) (Figure 6.2 E).

Another intriguing feature of ZIKV virions is their thermal stability. Kostyuchenko et al. (2016) compared the impact of temperature on the virus structures and infectivity of both ZIKV and DENV (6). In this study, ZIKV (H/PF/2013 strain) was grown in the mosquito cells at 28 °C, and purified at 4 °C by polyethylene glycol (PEG) precipitation, a sucrose cushion, followed by a potassium tartrate gradient. Then, the ZIKV samples were incubated at 28 °C, 37 °C, and 40 °C (mimicking high fever) for 30 minutes and then imaged by cryo-EM (6). At 28 °C, there were broken and shriveled particles with some smooth-surfaced particles with a diameter of about 50 nm. In contrast, there were many more smooth-surfaced particles at 37 °C and 40 °C.

It was speculated that the presence of shriveled particles at 28 °C could be due to the high osmolality during purification, and the higher temperature (37 °C or 40 °C) caused ZIKV virions to expend into smooth surfaced particles, making the lipid envelop more fluid and allowing the viral structure to revert to its normal state (6).

At 37 °C, similar structural changes were observed in some strains of DENV2 (New Guinea C, NGC) but not others (PVP94/07) (11–13). DENV4, which was more stable than DENV2 (NGC), remained as a smooth surfaced particle at 40 °C; however, considerable aggregation of virus particles formed (14). ZIKV seemed more stable compared to DENV4 since ZIKV particles did not form aggregation at 40 °C. Further plaque assay results confirmed that ZIKV retained its infectivity, whereas DENV2 and DENV4 showed significantly reduced infectivity at higher temperatures (6). The thermally stable structure of ZIKV may help it to survive in the harsh conditions of semen, saliva, and urine (15–17).

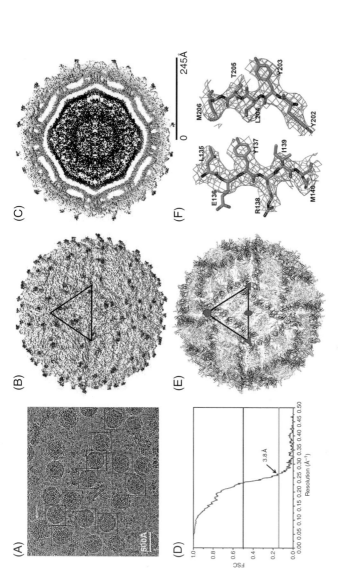

Figure 6.2 The cryo-EM structure of ZIKV at 3.8 Å. (A) A representative cryo-EM image of frozen, hydrated ZIKV, with smooth, mature virus particles highlighted with a surrounding black box, and a partially mature virus particle indicated by the yellow arrow. (B) A surface-shaded depth cued representation of ZIKV. The black triangle indicates the asymmetric unit. (C) A cross-section view of ZIKV displaying the radial density distribution. Panels B and C are color-coded based on the radii. Blue: up to 130 Å; cyan: 131 Å to 150 Å; green: 151 Å to 190 Å; yellow: 191 Å to 230 Å; red: from 231 Å. (D) A plot of the Fourier shell coefficient (FSC) showing the reconstruction resolution of 3.8 Å. (E) The Cα backbone of the E and M proteins in the icosahedral ZIKV particle showing the herringbone organization. The orientation is the same as in Panel B. E protein domains are color-coded based on standard designation. Red: domain I; yellow: domain II; blue: domain III. (F) Representative cryo-EM electron densities of several amino acids of the E protein. *Source:* Extracted from Sirohi et al. (2016) (10).

6.2   Binding and Entry

ZIKV infection begins with the virus binding to host cell receptors, followed by endocytosis (Figure 6.3). ZIKV receptors are not well characterized, perhaps because the virus has broad tissue tropism (e.g. brain, placenta) and may use a range of cellular receptors for different types of target cells (e.g. neuronal cells, trophoblast cells, macrophages) (18, 19), as many other flaviviruses do [reviewed in (20)]. A number of cell surface molecules have been implicated in ZIKV attachment and entry, including Dendritic cell-specific intercellular adhesion molecule-3-grabbing non-integrin (DC-SIGN), AXL, Tyro3, and, to a lesser extent, T-cell immunoglobin and mucin domain (TIM) (21).

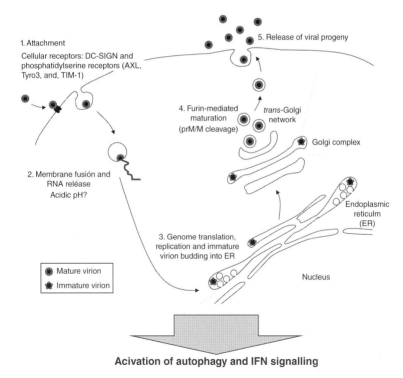

Figure 6.3 Overview of the ZIKV replication cycle. *Source:* Extracted from Saiz et al. (2016) (2).

C-type (calcium-dependent) lectin receptors are cellular proteins that bind mannose-rich glycans and are specialized in sensing invading pathogens (22, 23). DC-SIGN is a type 2 transmembrane C-type lectin, and is highly expressed on some macrophage subsets and immature dendritic cells (DCs) (24, 25). DC-SIGN has been reported to facilitate flavivirus binding and infection (20, 26–28). HEK293T cells are not susceptible to ZIKV infection; however, expression of DC-SIGN on these cells strongly enhances viral infection, demonstrating the importance of DC-SIGN in ZIKV entry (21).

Tyro3 and AXL receptors, members of TAM family, are protein tyrosine kinases (PTK) that regulate the clearance of apoptotic cells and inhibition of innate immunity (29–31). Tyro3 is primarily found in the central nervous system, while AXL is broadly expressed, and can be found in radial glia and blood vessels in developing cortex (32, 33). Ectopic expression of Tyro3 or AXL has been shown to highly enhance ZIKV infection in HEK293T cells (21). Human skin fibroblasts express AXL but not TIM-1; however, siRNA knock down of AXL effectively abrogates ZIKV infection (21). Many human primary cell lines (e.g., human radial glial cell, astrocyte, and endothelial cell) that express high level of AXL receptors are susceptible and vulnerable to ZIKV infection (33, 34). Loss-of-function study in human endothelial cells further reveals that an AXL receptor is required for virus entry at a postbinding step (34). Furthermore, herpes simplex virus-2 (HSV-2) infection enhances placental sensitivity to ZIKV by inducing the expression of TAM receptors (35).

The human TIMs (TIM-1, TIM-3, and TIM-4) share a type I cell surface glycoprotein structure: an extracellular N-terminal immunoglobulin-like domain, an O-linked-glycosylated mucin-like domain (MLD), and a transmembrane-spanning domain followed by a short cytoplasmic tail (36). Ectopic expression of TIM-1 and TIM-4 massively enhances infection of several flaviviruses, including all DENV serotypes, WNV, and YFV (37, 38). However, the expression of TIM-1 or TIM-4 has only modest or marginal effects on ZIKV entry in HEK293T cells (21). Although TIM contributes little to ZIKV infection, its expression shows an additive effect on the efficacy of AXL-mediated viral entry (21).

Subsequent to the early step of viral entry, late endosomal acidification triggers the fusion of host and viral membranes, permitting the entry of ZIKV positive sense RNA genome into the host cell cytosol. An early observation showed that ZIKV particles were sensitive to acidic pH, and were inactivated when treated with acidic pH lower than 6.2 (39). In addition, Delvecchio et al. (2016) showed that Chloroquine, a weak base that can rapidly increase pH in the acidic vesicles, significantly suppressed ZIKV replication in human neural stem cells and mouse neurospheres (40). A CRISPR/Cas9-based screening further confirmed that ZIKV infection is through the endocytic pathway. Several proteins involved in endocytosis, such as RAB5C, RABGEF1, WDR7, and ZFYVE20, are required for ZIKV replication (41).

6.3 Genome Structure

Following uncoating of the nucleocapsid, viral genome is released inside the cytoplasm. The positive-strand viral genome has distinct roles during the life cycle, serving as the mRNA to initiate translation of viral proteins and replication of viral genome, as a template for negative-strand RNA synthesis, and as the genetic materials packaged into newly assembled virions. ZIKV has a positive-sense, single-stranded RNA genome of 10,974 nucleotides long (strain MR-766) (42). The viral genome is composed of 5′-capped untranslated region (UTR), a continuous open reading frame (ORF) encoding a 3,419 amino acids (aa) long polypeptide (42), which is processed into three structural proteins, the capsid protein (C), the precursor of membrane protein (prM), and the envelope protein (E), as well as seven nonstructural proteins, NS1, NS2A, NS2B, NS3, NS4A, NS4B, and NS5, and a 3′-UTR essential for viral replication (Figure 6.4) (42).

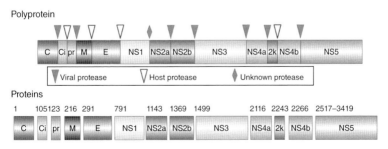

Figure 6.4 ZIKV polyprotein processing to produce mature peptides. *Source:* Extracted from Sun et al. (2017) (79).

ZIKV genomic RNA contains a number of important modifications, including 5′ cap structure, N^6 methylation of adenosine (m^6A), and 2′-O ribose methylation (2′-O-Me) (43). All four nucleotides contain 2′-O-Me groups, with U and C representing the most extensive modifications, followed by A and G (43).

m^6A is the most prevalent internal modification of the eukaryotic mRNA (44–46). m^6A nucleosides are abundant in ZIKV RNA, containing 12 m^6A peaks across the viral genome (43, 47). Adenosine methylation in ZIKV RNA is controlled by host methyltransferases METTL3 and METTL14, as well as demethylases ALKBH5 and FTO. Knockdown of methyltransferases increases while silencing demethylases decreases ZIKV replication (43). Furthermore, members of YTH domain protein family (YTHDF), including YTHDF1, YTHDF2, and YTHDF3, are able to bind to methylated ZIKV RNA and regulate the stability of viral RNA (43). Such regulation is also found during HCV infection (47).

Like other flaviviruses, ZIKV produces noncoding subgenomic flaviviral RNAs (sfRNAs) (48–50). sfRNAs are not transcribed from the viral genome but, rather, accumulate as a result of incomplete degradation of viral genomic RNA by host exonucleases, such as Xrn1 (48, 51). These sfRNAs regulate viral replication and translation, and contribute to their pathogenicity (52); they dysregulate RNA decay pathways and counteract antiviral innate immunity (53–57). It has been shown that a pair of stable Xrn1-resistent RNA structures, xrRNA1 and xrRNA2, at the 5′ end of 3′ UTR of flaviviruses are responsible for the formation of sfRNAs (48).

The RNA structure, which is the source of ZIKV sfRNAs, forms a complex fold and is stabilized by pseudoknots (49). Pseudoknots are tertiary RNA structures that form as a result of base pairing of free nucleotides in the loop of a hairpin structure to a complementary sequence elsewhere in the RNA (48, 58, 59). In the xrRNA1 of ZIKV, the 5′ end of the RNA sequence is nestled within a 14-nucleotide ring made from a junction of three helices. The xrRNA1 structural integrity is further strengthened by long-range nucleotide interactions, including a pseudoknot formed from four constative G-C pairs (49, 51). Mutations in xrRNA1 that disrupt the pseudoknot structure or the three-way junction result in significant reduced formation of sfRNAs during virus infection (49). A recent study has revealed that the ZIKV sfRNAs function as antagonists of RIG-I-mediated and to a lesser extent of MDA-5-mediated induction of type I interferon (IFN)

(50). The presence of an evolutionarily conserved viral RNA structure that leads to the formation of sfRNAs suggests that an essential function of the sfRNAs for productive flavivirus infection (51).

Interestingly, it has recently been identified that ZIKV viral genome potentially harbors G-quadruplex sequences (PQSs) (60). G-quadruplex folds occur when there are at least four contiguous runs of two or more guanosine (G) nucleotides in a short sequence. The Gs fold around cellular K^+ ions to form G-tetrads composed of four Hoogsteen base-paired Gs (61). DNA PQSs are able to function as critical cis-acting regulatory elements in many signaling pathways, while RNA PQSs have been ascribed with functions related to mRNA splicing, transcriptional termination, and translational control (62, 63). A study using over 56 members of flaviviruses, including ZIKV, identified seven conserved PQSs, which were found in the coding regions for the prM, E, NS1, NS3, and NS5 proteins (60). However, this inspection failed to identify any PQS in the negative-sense strand of ZIKV. Although the functional purposes for PQSs have been observed in the human immunodeficiency virus (HIV), Epstein-Barr virus (EBV), herpes simplex virus (HSV), and hepatitis C virus (HCV) (64–67), the role of PQSs in ZIKV infection is still unclear.

6.4 Translation and Proteolytic Processing

Protein translation is composed of an intricate series of events. This process requires many components that viral genomes cannot encode. Therefore, viruses fully rely on the translation machinery of host cells to produce the viral proteins and, as a result, have developed novel mechanisms to compete with cellular mRNAs for limiting translation factors [reviewed in (68)].

The efficiency of viral genome translation is a major determinant of flavivirus infectivity (69). Following the release of viral genomic RNA, a polyprotein precursor of approximately 3,400 amino acids in length is translated. The viruses have developed numerous mechanisms to hijack the host cell translation apparatus in order to ensure translational competence. The existence of $5'$ m^7GpppN-capped structure and $3'$ non-polyadenylated untranslated region (UTR) indicates that the viruses undergo translation in a cap-dependent manner. However, flavivirus genome translation may switch from a canonical cap-dependent to cap-independent manner under conditions of translation suppression (70, 71). Flaviviruses, such as DENV and ZIKV, uncouple translation suppression from cellular stress response through inhibiting the phosphorylation of the translation initiation factor eIF2α and repressing the formation of stress granules (71). 2′-O methylation of the 5′ cap structure subverts the antiviral effects of IFN-induced proteins with tetratricopeptide repeats (IFIT), which are IFN-stimulated genes (ISGs) implicated in the regulation of protein translation (72).

The polyprotein precursor is co- and post-translationally processed into three structural and seven nonstructural (NS) proteins by a combination of cellular proteases (furin-type and other Golgi-localized proteases), and the viral serine protease NS3Pro/NS2B and the N-terminal domain of nonstructural protein 3 (NS3Pro), which requires NS2B for its activity. Host signal peptidase is responsible for cleavages of Ci (intracellular C)/prM, prM/E, E/NS1, 2k/NS4B (73, 74), and also likely NS1-NS2A (75) while the host Golgi protease furin potentially cleaves the pr/M junction (76). Virus-encoded NS3Pro/NS2B processes the remaining peptide bonds at the C/Ci, NS2A/NS2B, NS2B/NS3, NS3/NS4A, NS4A/2 K, and NS4B/NS5 junctions (Figure 6.4) (77–79). All these mature viral proteins control virus biology, including virus replication, transmission, pathogenicity, and host immunologic responses, thus influencing disease prognosis. Currently, cleavages of all ZIKV mature peptides are only speculated based on the multiple sequence alignments with other flaviviruses, and have not yet been supported by experimental sequencing evidence (79).

6.5 Features of the Nonstructural Proteins

Currently, there are very limited studies addressing the functions of nonstructural proteins of ZIKV but some could be inferred from related flaviviruses (1, 80, 81).

6.5.1 NS1 Glycoprotein

ZIKV NS1 glycoprotein is a ~48 kDa conserved nonstructural protein with two conserved N-linked glycosylation sites and six intramolecular disulfide bonds (Figure 6.5 B) (82–84). NS1 is translocated into the ER during protein synthesis, and could form monomers, dimers, or hexamers following post-translational modifications (85–87). Intracellular form of NS1 is localized to sites of viral RNA synthesis

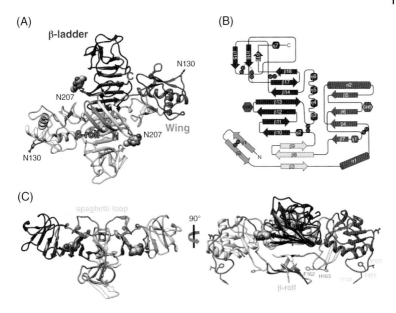

Figure 6.5 Structure of ZIKV NS1 dimer. (A) Ribbon diagram of NS1 dimer, with one subunit in gray and the other color-coded by domain. Green: β-roll; orange: wing; magenta: connector subdomain; blue: central β-ladder. N-linked glycosylation sites are indicated in sticks, and the glycans are shown in spheres. (B) Topology diagram of the NS1 monomer. Color pattern is the same as in Panel A. Glycosylation sites are shown as red hexagons and disulfide bonds as yellow circles. (C) Side views of the NS1 dimer from the wing (left panel) and the end of the β-ladder (right panel). The β-roll, connector subdomain, and the intertwined loop (yellow) of the wing domain form a discontinuous protrusion on one face of the β-ladder with the spaghetti loop (cyan) on the other face. The wing domain is omitted from the left image for clarity. Color pattern is the same as in Panel A. *Source:* Extracted from Xu et al. (2016) (82). (*See insert for color representation of the figure.*)

and plays a key role in genome replication (88, 89). NS1 could be secreted into the extracellular space as hexameric lipoprotein particles of 10 nm appearing as three dimers held together in a barrel configuration (87, 90). It is speculated that ER membrane-associated NS1 dimer plays an essential organizational role in the formation of the replication complex through interaction with NS4A and NS4B (88, 91, 92). Although not presenting in the virion, NS1 is the major antigenic marker for viral infection, and host immune recognition and evasion (93, 94). Secreted NS1 could bind to complement proteins and modify or antagonize their functions (94–96). In addition,

NS1 was also shown to activate Toll-like receptor 4 signaling in primary human myeloid cells, resulting in the secretion of pro-inflammatory cytokines and vascular leakage (97, 98).

Recently, several studies have revealed the structure of ZIKV NS1 (82–84). The overall structure of full-length ZIKV NS1 (BeH819015 strain, Brazil) is very similar to WNV and DENV2 NS1 structures. Each NS1 monomer has three domains: β-hairpin ("β-roll") domain, β-ladder domain, and wing domain (Figure 6.5 A) (82). Two NS1 are assembled into a symmetric, head-to-head dimer, which is organized around the β-roll domain and β-sheets extended from the β-ladder domain.

Wang et al. (2013) showed that a pathogenic mutation T233A, which is found in the brain tissue of infected fetus with neonatal microcephaly, destabilized the NS1 dimeric assembly *in vitro* (83). Thr233, a unique residue found in ZIKV but not in other flaviviruses, is located at the dimer interface. Mutation of Thr233 to Ala disrupted the elaborated hydrogen bonding network at NS1 dimer interface (83).

In the ZIKV full-length structure, the long loop, consisting of Y122, F123, and V124, forms a hydrophobic "spike" that may contribute to membrane association (Figure 6.5 C) (82). This loop is critical in mediating NS1 interaction with the envelope protein, thus tethering the replication complex with virion formation (99). In the modeled ZIKV NS1 hexamer, the intertwined loop "spike" of one monomer is very close to the β-ladder domain of the neighboring NS1 monomer, indicating that the intertwined loop might participate in the NS1 hexamer formation (Figure 6.6) (82).

6.5.2 NS2A and NS2B Proteins

NS2A is a relatively small (~22 kDa) hydrophobic, membrane-spinning protein (100). NS2A is a component of the viral replication complex that functions in virion assembly and antagonizes the host immune response in six ways:

1) NS2A is important for viral RNA synthesis. Kunjin virus (KUNV) NS2A localizes to subcellular sites of RNA replication and interacts with the 3′ UTR, NS3, and NS5 in the form of a replication complex (101).
2) NS2A functions in the virus assembly. Mutations in NS2A of YFV, KUNV, and DENV2 block virus maturation/assembly (100, 102, 103).

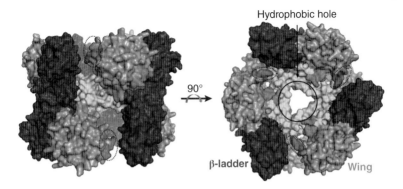

Figure 6.6 Model of the ZIKV NS1 hexamer. The ZIKV NS1 hexamer is modeled using the full-length DENV2 NS1 structure (PDB code: 4O6B) (236). Each monomer is color-coded by domains. Green: β-roll; orange: wing; magenta: connector subdomain; blue: central β-ladder. The intertwined loop "spike" of ZIKV NS1 involved membrane association is shown in red and highlighted by dotted lines. *Source:* Extracted from Xu et al. (2016) (82). (*See insert for color representation of the figure.*)

3) DENV2, WNV, and KUNV NS2A can inhibit IFN signaling because specific mutations in the protein diminish this inhibitory activity (104–106). JEV NS2A is capable of blocking dsRNA-activated protein kinase PKR (107).
4) YFV serine protease can cleave an internal site in NS2A to generate a C-terminally truncated form, NS2Aα (108, 109).
5) In the JEV serogroup, a conserved slippery heptanucleotide motif, which is located at the beginning of the NS2A gene, is required for the production of NS1 through a ribosomal frameshift mechanism (110).
6) NS2A participates in virus-induced membrane formation during virus assembly (103).

NS2B is a small (∼14 kDa) membrane-associated protein (111). NS2B, associated with NS3, forms a stable NS2A-NS3 complex, possessing serine protease activity. In this complex, NS2B serves to anchor this complex to cellular membranes and acts as an essential co-factor for the NS2B-3 serine protease (112). The minimal cofactor region of the NS2B comprises the hydrophilic residues 49–97 (77), with N-terminal 18 residues (49–67) supporting proper fold of the NS3 protease (113, 114) and C-terminal part (68–96) helping create a substrate binding site of the protease (114–116).

6.5.3 NS3 Protein

The NS3 protein is a large (~70 kDa), multifunctional protein, a key component for viral polypeptide processing and genomic replication, containing a catalytic protease domain of NS2B-NS3 serine protease at its N-terminus and a helicase domain at the C-terminus (1). The NS2B-NS3 serine protease functions to cleave the virus peptides and has specificity for substrates containing adjacent basic residues at the C/Ci, NS2A/NS2B, NS2B/NS3, NS3/NS4A, NS4A/2 K, and NS4B/NS5 junctions (117–120). A structural study of unlinked ZIKV NS2B-NS3 protease has shown that free protease predominantly adopts a closed conformation while local conformational dynamics occurs at the NS2B-NS3 binding interface (121). Peptide binding of the substrate does not further induce significant conformational changes (121).

Similar to other flaviviruses, the C terminal region of NS3 encodes a superfamily 2 (SF2) RNA-helicase-nucleoside triphosphtase (NTPase) (122). The tertiary structure of ZIKV NS3 helicase is composed of three domains of roughly similar size (Figure 6.7 B) (123, 124). Overall, all three domains superimpose very well on equivalent domains in other flaviviruses (123).

Domain I (residue 175–332) and II (residue 333–481) share a tandem α/β RecA-like fold, although there is little sequence identity between these two domains. Domain I contains the classical motifs I (P loop or Walker A), Ia, II (or Walker B), and III, while domain II contains motifs IV, IVa, V, and VI. The NTPase active site is located in the cleft between domains I and II, in which motifs I and II play an important role in recognizing NTP and cations (Mn^{2+} and Mg^{2+}) (Figure 6.7 C) (123–125). Besides, this region also exhibits RNA triphosphatase (RTPase) activity, which is proposed to dephosphorylate the 5′ end of the genome before cap addition (126). RTPase site appears to overlap with NTPase active site that powers the helicase (127, 128). Domain III (residue 482–617) is predominantly α-helical. In the structure, a positively charged tunnel can be identified along the boundary of domain III. The tunnel is lined with positively charged residues and remains wide enough to accommodate a ssRNA in an extended conformation running through domain II to I (Figure 6.7 C) (123, 124). In addition, this domain has also been implicated in interaction with the RNA-dependent RNA polymerase NS5 in other flaviviruses (129). Lastly, Langat virus (LGTV), DENV2, and WNV NS3 have been shown to induce apoptosis, in some cases

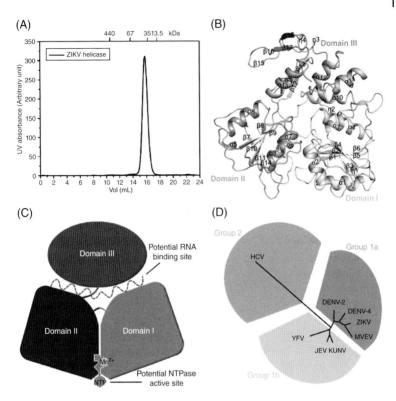

Figure 6.7 The structure of ZIKV NS3 helicase. (A) Size-exclusion chromatograms of ZIKV helicase. (B) Ribbon diagram of ZIKV helicase structure. (C) A cartoon diagram showing the overall fold with potential RNA binding site and NTPase active site highlighted. (D) Structure-based phylogenetic tree diagram of eight viral helicases from the Flaviviradae family using the program SHP and PHYLIP (237, 238). The PDB IDs of viral helicase structures are listed as follow: HCV (1HEI), DENV-2 (2BMF), DENV-4 (2JLQ), MVEV (2V8O), KUNV (2QEQ), JEV (2Z83), and YFV (1YKS). *Source:* Extracted from Tian et al. (2016) (124).

through activation of caspase-8 (130–132), while DENV2 NS2–NS3 protease can also inhibit the activation of type I IFN pathway in human dendritic cells (133).

6.5.4 NS4A and NS4B Proteins

NS4A and NS4B are small hydrophobic proteins with molecular weight of 16 kDa and 27 kDa, respectively. These two proteins are essential components of the ER membrane-associated replication complex (134). NS4A and NS4B are linked by a conserved 23-amino-acid-long signal peptide 2 K. Similar to the coordinated processing of C protein, the cleavage between 2 K and NS4B by NS2B/NS3 protein-ase requires prior cleavage between NS4A and 2 K by a host signal peptidase (120). DENV2 NS4A consists of 127 amino acids and forms two transmembrane domains (TMDs) while NS4B consists of 248 amino acids and forms three TMDs (135, 136). NS4A and NS4B have at least seven functions in viral replication (134):

1) NS4A and 2 K play a regulatory role in ER membrane rearrange-ment. Deletion of KUNV 2 K leads to redistribution of NS4A to the Golgi apparatus (137). However, expression of DENV NS4A lack-ing the 2 K fragment results in ER membrane alterations resem-bling virus-induced structures, whereas expression of full-length NS4A does not induce comparable membrane alterations (135). All these results suggest that the function of NS4A in modulating the ER membrane is regulated by 2 K through distinct mechanisms in different flaviviruses.

2) Interaction between DENV NS4A and cellular vimentin intermedi-ate filaments regulates virus replication complex formation (138).

3) Flavivirus NS4A and NS4B proteins induce autophagy. ZIKV NS4A and NS4B have been shown to induce autophagy and inhibit neuro-genesis through deregulating Akt-mTOR signaling pathway in human fetal neural stem cells (139). DENV NS4A alone induces autophagy to prevent cell death and facilitate viral replication (140).

4) WNV NS4A regulates the ATPase activity of the NS3 helicase (141), while DENV NS4B dissociates NS3 from ssRNA and conse-quently modulates the helicase activity of NS3 (142, 143).

5) NS4A and NS4B genetically interact with NS1 to modulate viral replication. The replication defect of YFV-bearing NS1 mutations could be rescued by an adaptive mutation in viral NS4A (88), while the replication defect of WNV containing NS1 mutations could be compensated by a mutation in viral NS4B (92).

6) A replication defect caused by an N-terminal cytoplasmic muta-tion in DENV NS4A can be rescued by mutations in the TMD3 of NS4B (144).

7) Both NS4A and NS4B proteins are able to inhibit IFN signaling (106, 145, 146), suggesting that NS4A and NS4B may function cooperatively in viral replication and host response though the direct interaction between NS4A and NS4B remains to be established and characterized (134).

6.5.5 NS5 Protein

NS5 is a large (~103 kDa), highly conserved, multifunctional phosphoprotein with methyltransferase (Mtase) and RNA-dependent RNA polymerase (RdRp) activities encoded within its N- and C-terminal regions, respectively. Formation of a type 1 RNA cap involves multiple steps, which requires both NS3 RTPase and NS5 MTase activities [reviewed in (147)]. The NS5 MTase methylates guanosine N-7 and ribose 2′-O positions of the viral RNA cap structure to facilitate translation of viral polypeptide and to inhibit the host innate immune response (72, 148, 149). The RdRp initiates RNA synthesis by a *de novo* initiation mechanism wherein a single-nucleotide triphosphate serves as a primer for nucleotide polymerization (150–152).

Crystal structure of ZIKV NS5 revealed that the MTase domain is dominated by a Rossmann fold with a seven-stranded β-sheet sandwiched by two α-helices from one side and another α-helix from the other side (153–158). Residues that bind GTP, catalyze methyl transfer, and bind the methyl donor S-adenosine-methionine (SAM) are conserved and arranged in a line within a concave surface of the MTase (Figure 6.8). Similar to other viral RdRps (159), the ZIKV NS5 RdRp adopts a classic "right-handed" structure consisting of three subdomains: Palm, Fingers, and Thumb (Figure 6.9) (153–156, 160). The RdRp domain also harbors two zinc ions, located in the Fingers subdomain, and at the junction of the Palm and Thumb subdomains (Figure 6.9). There are extensive interactions between the Fingers and Thumb subdomains to encircle the active site of the polymerase (Figure 6.9). The overall structure of the ZIKV NS5 has striking similarities to that of JEV NS5 while ZIKV MTase is in a distinct orientation in relative to the DENV MTase (153, 154), raising a question about the functional implication of these two conformation states of NS5 proteins, which requires additional biochemical and cellular analyses.

Previous mutagenesis studies showed that the methylation events are separable; N7 methylation of structure is required for viral translation and replication, whereas 2′-O methylation enables the virus to evade the innate immune defense (72, 161–163). Phosphorylation of YFV NS5 at Ser56, which is near the methyltransferase active site, by cellular casein kinase 1 dramatically inhibits 2′-O methylation and RNA replication (164, 165). In addition to its ability to initiate *de novo* RNA synthesis, NS5 also forms a complex with NS3 and stimulates NS3 NTPase and RTPase activities (166–169). Although WNV and

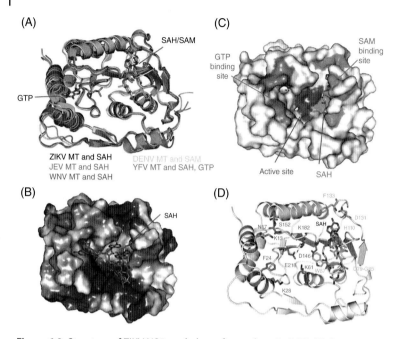

Figure 6.8 Structure of ZIKV NS5 methyltransferase domain (MT). (A) Comparison of MT structures of five flaviviruses. The PDB IDs of viral MT structures are listed as follow: DENV (3P97), YFV (3EVC), WNV (2OY0), and JEV (4K6M). SAH or SAM and GTP are shown in sticks. (B) Surface of ZIKV NS5 MT involved in Cap-0 RNA (5'-^{m7}G$_{0ppp}$A$_1$G$_2$U$_3$U$_4$G$_5$U$_6$U$_7$-3') binding. Positively and negatively charged surface are colored blue and red, respectively. (C) Surface view of ZIKV NS5 MT with colored active site (magenta), and binding sites for GTP (orange) and SAM (green). (D) Key residues of ZIKV NS5 MT essential for GTP binding (orange), SAM binding (green) and catalysis (magenta). SAH is shown in blue sticks. *Source:* Extracted from Zhao et al. (2017) (153). (*See insert for color representation of the figure.*)

DENV2 NS5 is localized at the site of viral RNA synthesis, this protein is frequently localized to the nucleus in the flavivirus-infected cells (167, 170–172), indicating additional roles of NS5 in virus life cycle. NS5 protein has been shown to inhibit the Jak-Stat2 pathway of type I IFN signaling (173–177).

6.6 RNA Replication

Following several rounds of translation, the flavivirus NS proteins presumably recruit the viral genome out of translation machinery and into a replication complex (1). RNA replication starts with the

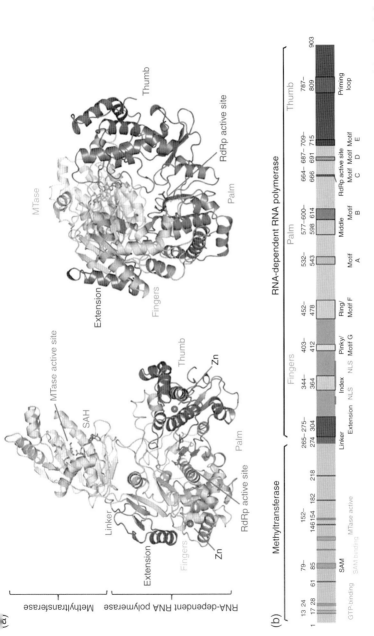

Figure 6.9 Structure of full-length ZIKV NS5. (a) Structure of full-length ZIKV NS5. A top view look into the active site of the RdRp (left panel) and a side view (right panel) are shown. (b) Schematic representation of ZIKV NS5 showing the locations of key residues and structural motifs. Panel A and B are color-coded accordingly based on domains and structural motifs. The active site residues of the MT and the RdRp are shown in pink and purple sticks, respectively. The SAH molecule binding to the MT is shown by the magenta stick model. *Source:* Extracted from Zhao et al. (2017) (153). (*See insert for color representation of the figure.*)

synthesis of a genome-length negative-strand viral RNA, which then serves as a template to direct positive-strand genomic RNA synthesis (1). Virus-infected cells undergo a massive remodeling of the endoplasmic reticulum (ER) forming membranous replication factories (178). As other flaviviruses, ZIKV infection induces invagination of the ER, leading to the appearance of smooth vesicular structures adjacent to ER-derived convoluted membranes (Figure 6.10) (171, 179–184). These structures, sometimes referred to as vesicle packets (VPs), are the presumed sites of viral RNA replication (171, 181). Such a spatial segregation structure would allow concentration of metabolites for RNA replication while protecting viral RNA from cellular nuclease and cytosolic RNA sensors. Time-course analyses revealed that virus-induced vesicles formed at as early as 4-hour post-infection (h.p.i.) and 8 h.p.i. for the MR766 and H/PF/2013 strains, respectively, and accumulated through the course of infection (184).

The average diameters of ZIKV-induced vesicles in hNPCs were approximately 62.80 nm and 65.51 nm for African and Asian strains, respectively, which were smaller compared to those in Huh7 cells, indicating that vesicle size might be cell-type dependent (184). Electron tomography revealed that these vesicles retained connectivity to the cytosol through a narrow pore-like structure, sometimes apposed to sites of virus assembly (Figure 6.10 B) (171, 184). Quantification of pore sizes in H/PF/2013- and MR766-infected hNPCs revealed an average width of approximately 11.4 nm and 10.7 nm, respectively, which were comparable to the pore size observed in DENV-infected cells (11.2 nm) (Figure 6.10) (171, 184).

Using immunofluorescence-based detection of dsRNA, ZIKV replication was detected at as early as 4 h.p.i., which is consistent with previous observation that negative-strand RNA synthesis started at as early as 3 h.p.i. (184, 185). Viral RNA synthesis is asymmetric, with approximately 10-fold more positive-strand RNA accumulating than negative-strand viral RNA (186, 187). During the viral replication process, three major species of viral RNA are required:

1) The positive-stranded genomic RNA
2) A double-stranded replicative form (RF), which is RNase resistant and can be used as a recycling template for semiconservative replication
3) A heterogeneous population of replicative intermediates (RIs) that most likely represent duplex regions and newly synthesized RNAs of nascent strands undergoing elongation (1, 186, 188).

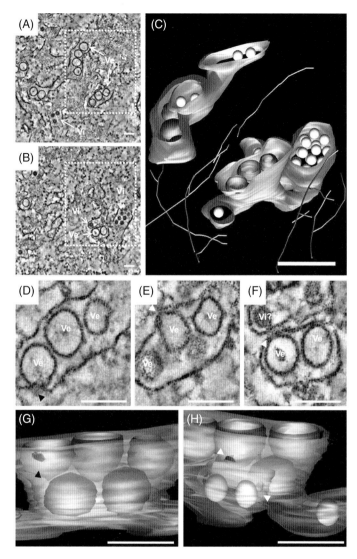

Figure 6.10 Electron tomography of ZIKV-induced vesicle packets in hNPCs. (A and B) Electron tomography analysis showing ZIKV-induced vesicles (Ve) within the rough ER and virions (Vi) in hNPC. hNPCs were fixed at 24 h.p.i., and embedded in epoxy resin for electron tomography analysis. (C) 3D surface model of the boxed area in panel A, showing virus-induced vesicles (dark gray), virions (white), and intermediate filaments (lines). ER membranes are shown in light blue. (D–F) Slice through the tomogram showing the pore-like openings (arrowheads) of ZIKV-induced vesicles toward the cytoplasm. A potential ZIKV budding event (Vi?) on the ER tubule opposing the vesicle pore was observed (F). (G and H) Reconstruction of the areas shown in panel D and E. Arrowheads refer to the vesicle pores marked in panel D and E. Scale bars, 100 nm in (A), (B), and (D–H); 200 nm in (C). *Source:* Extracted from Cortese et al. (2017) (184).

Pulse-chase analyses indicate that RF and RI are precursors to genomic RNA (1, 186, 188).

6.7 Features of the Structural Proteins

6.7.1 Capsid Protein

Capsid (C) protein is a highly basic protein of 11 kDa. Its C-terminal hydrophobic tail serves as a signal peptide for ER translocation of prM and is cleaved by viral NS2B-NS3 serine protease. The mature C protein tends to fold into a dimer with each monomer containing four α-helices (α1 to α4) and an unstructured N-terminal region (189–191). Based on the charge distribution on the dimer surface, the unstructured N-terminal region and the charged residues at the C-terminus (α4–α4′) are involved in viral RNA binding, while on the opposite side, a large uncharged conserved surface formed by α1–α1′ and α2–α2′ is proposed to interact with viral envelope and biological membranes (Figure 6.11 B and C) (191–193). In addition to its structural function of protecting the viral genome, C protein has four multifunctional properties:

1) C protein is very tolerant for large deletions without losing RNA packaging capability. YFV and TBEV C proteins with a large number of deletions retain their abilities to package viral RNA; however, internal deletions of the hydrophobic residues are less permissive (194, 195).
2) A number of studies have reported the presence of C proteins in the nucleus of flavivirus-infected cells (196–198). In the nucleus, Flavivirus C proteins may interact with core histones and disrupt nucleosome formation (199). It was also reported that C protein can interact with other nucleus proteins, including hnRNP-K, Daxx, and nucleolin though the role of C protein is still elusive (200–202).
3) Flavivirus C proteins have been shown to associate with several deleterious responses to the host cells. Flavivirus C protein can induce cytotoxic effect in the virus-infected cells via activation of caspase-9 (203) or Fas-mediated activities (201, 202).
4) Despite poor amino acid homology, flavivirus C proteins show conserved physical-chemical properties in terms of high net positive charge and relative molecular weight (189, 191, 204).

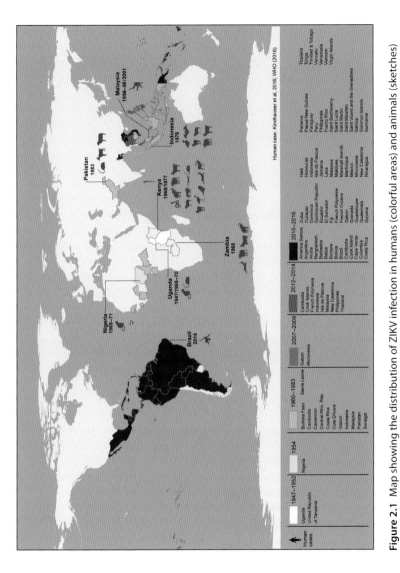

Figure 2.1 Map showing the distribution of ZIKV infection in humans (colorful areas) and animals (sketches) following the chronological outbreaks in different regions. *Source:* Figure and legend extracted from Bueno et al. (2016) (68).

Zika Virus and Diseases: From Molecular Biology to Epidemiology, First Edition.
Suzane Ramos da Silva, Fan Cheng and Shou-Jiang Gao.
© 2018 John Wiley & Sons, Inc. Published 2018 by John Wiley & Sons, Inc.

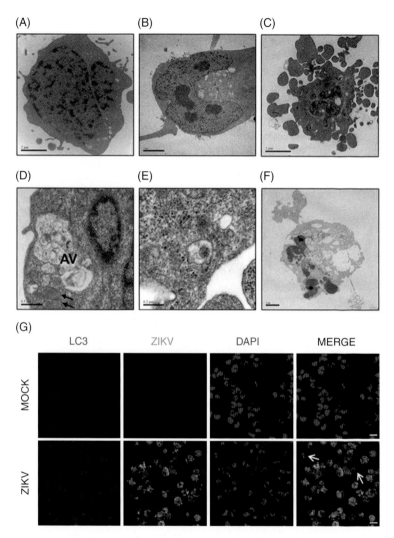

Figure 4.2 (A, B) Transmission electron micrographs of mock-infected cells, showing nuclei and organelles with normal aspect after 24 and 72 hours of culture. (C–F) ZIKV-infected cells 24 hours (C, D), 48 hours (E) and 72 (F) hours after infection. Cells in early (C) and late (F) apoptotic processes. (D) The presence of large perinuclear autophagic vacuoles (AV) can be observed. Black arrows indicate mitochondria with altered morphology. (E) Presence of viral capsid in intracellular vacuole. (G) Immunostaining for ZIKV (green) and the autophagic vacuole marker LC3 (red). Nuclei are stained with DAPI (blue). White arrows indicate cells with negative or low ZIKV staining, and absence of perinuclear LC3 staining. Scale bars = 20 μm. *Source:* Figure and legend extracted from Souza et al. (2016) (73).

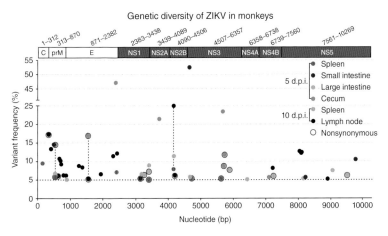

Figure 5.3 Analysis of ZIKV diversity in different tissues from monkey rhesus. Dotted lines indicate consensus changes. *Source:* Figure and legends are adapted from Li et al. (2016)(26).

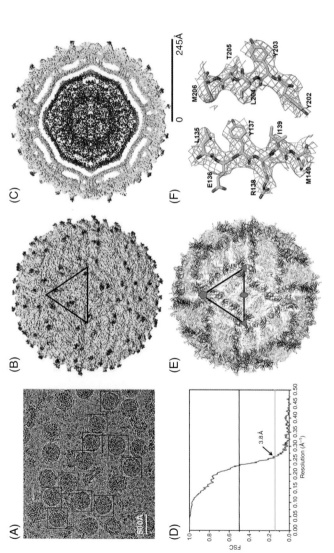

Figure 6.2 The cryo-EM structure of ZIKV at 3.8 Å. (A) A representative cryo-EM image of frozen, hydrated ZIKV, with smooth, mature virus particles highlighted with a surrounding black box, and a partially mature virus particle indicated by the yellow arrow. (B) A surfaceshaded depth cued representation of ZIKV. The black triangle indicates the asymmetric unit. (C) A cross-section view of ZIKV displaying the radial density distribution. Panels B and C are color-coded based on the radii. Blue: up to 130 Å; cyan: 131 Å to 150 Å; green: 151 Å to 190 Å; yellow: 191 Å to 230 Å, red: from 231 Å. (D) A plot of the Fourier shell coefficient (FSC) showing the reconstruction resolution of 3.8 Å. (E) The Cα backbone of the E and M proteins in the icosahedral ZIKV particle showing the herringbone organization. The orientation is the same as in Panel B. E protein domains are color-coded based on standard designation. Red: domain I; yellow: domain II; blue: domain III.

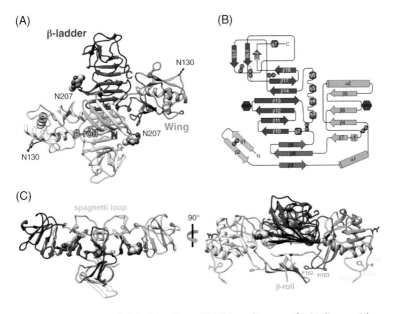

Figure 6.5 Structure of ZIKV NS1 dimer. (A) Ribbon diagram of NS1 dimer, with one subunit in gray and the other color-coded by domain. Green: β-roll; orange: wing; magenta: connector subdomain; blue: central β-ladder. N-linked glycosylation sites are indicated in sticks, and the glycans are shown in spheres. (B) Topology diagram of the NS1 monomer. Color pattern is the same as in Panel A. Glycosylation sites are shown as red hexagons and disulfide bonds as yellow circles. (C) Side views of the NS1 dimer from the wing (left panel) and the end of the β-ladder (right panel). The β-roll, connector subdomain, and the intertwined loop (yellow) of the wing domain form a discontinuous protrusion on one face of the β-ladder with the spaghetti loop (cyan) on the other face. The wing domain is omitted from the left image for clarity. Color pattern is the same as in Panel A. *Source:* Extracted from Xu et al. (2016) (82).

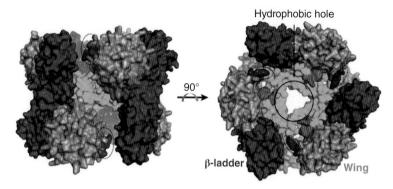

Figure 6.6 Model of the ZIKV NS1 hexamer. The ZIKV NS1 hexamer is modeled using the full-length DENV2 NS1 structure (PDB code: 4O6B) (236). Each monomer is color-coded by domains. Green: β-roll; orange: wing; magenta: connector subdomain; blue: central β-ladder. The intertwined loop "spike" of ZIKV NS1 involved membrane association is shown in red and highlighted by dotted lines. *Source:* Extracted from Xu et al. (2016) (82).

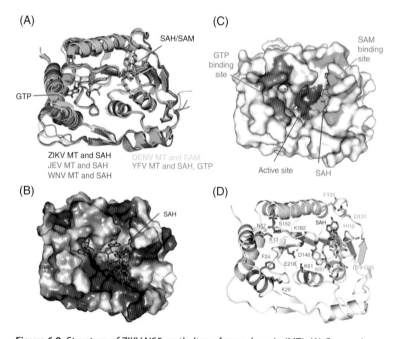

Figure 6.8 Structure of ZIKV NS5 methyltransferase domain (MT). (A) Comparison of MT structures of five flaviviruses. The PDB IDs of viral MT structures are listed as follow: DENV (3P97), YFV (3EVC), WNV (2OY0), and JEV (4K6M). SAH or SAM and GTP are shown in sticks. (B) Surface of ZIKV NS5 MT involved in Cap-0 RNA ($5'$-$^{m7}G_{0ppp}A_1G_2U_3U_4G_5U_6U_7$-$3'$) binding. Positively and negatively charged surface are colored blue and red, respectively. (C) Surface view of ZIKV NS5 MT with colored active site (magenta), and binding sites for GTP (orange) and SAM (green). (D) Key residues of ZIKV NS5 MT essential for GTP binding (orange), SAM binding (green) and catalysis (magenta). SAH is shown in blue sticks. *Source:* Extracted from Zhao et al. (2017) (153).

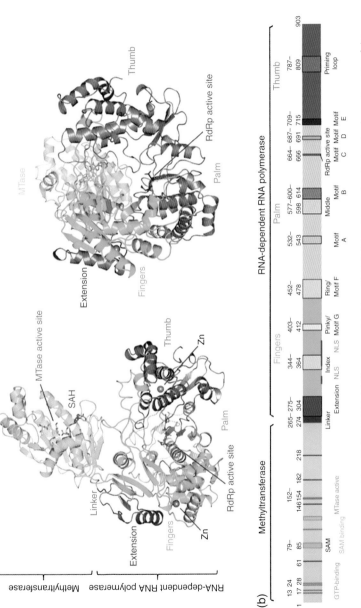

Figure 6.9 Structure of full-length ZIKV NS5. (a) Structure of full-length ZIKV NS5. A top view look into the active site of the RdRp (left panel) and a side view (right panel) are shown. (b) Schematic representation of ZIKV NS5 showing the locations of key residues and structural motifs. Panel A and B are color-coded accordingly based on domains and structural motifs. The active site residues of the MT and the RdRp are shown in pink and purple sticks, respectively. The SAH molecule binding to the MT is shown by the magenta stick model. *Source:* Extracted from Zhao et al. (2017) (153).

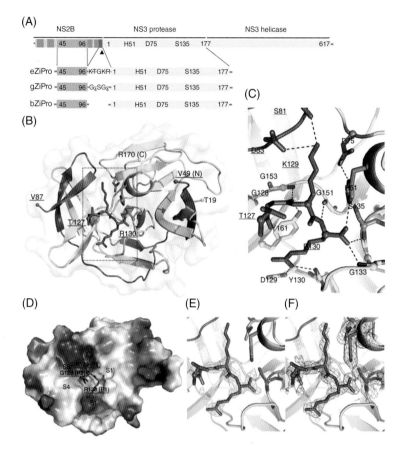

Figure 9.4 Crystal structure of eZiPro in complex with the C terminal TGKR tetrapeptide of NS2B. (A) Full-length NS2B and NS3 proteins and construct designs of eZiPro, gZiPro, and bZiPro. (B) Overall structure of eZiPro showing the TGKR NS2B peptide bound in substrate binding site. NS2B and NS3 are colored in magenta and yellow, respectively. (C) Interactions between viral peptide and residues from protease. (D) Surface charge density view of NS2B-NS3 complex. Residues of substrate binding pockets are labeled. (E) A simulated annealing omit map of the TGKR peptide is contoured at 3σ in green mesh. (F) 2mFo-DFc electron density map contoured at 1σ in blue. *Source:* Extracted from Phoo et al. (2016) (66).

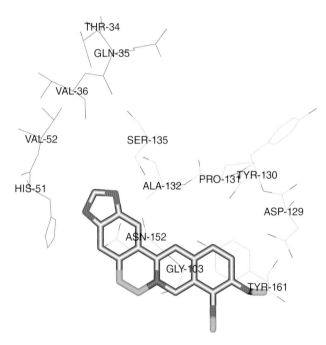

Figure 9.7 Molecular docking complex of berberine with nonstructural 3 (NS3) protein of Zika virus. *Source:* Extracted from Sahoo et al. (2016) (73).

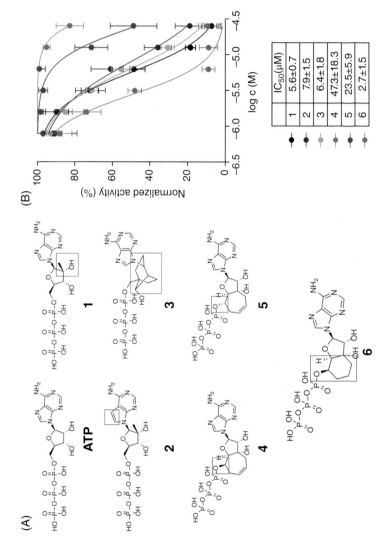

Figure 9.9 *In vitro* activity of ATP analogs. (A) Structural formula of ATP and its analogs. (B) *In vitro* activity of selected triphosphates. *Source:* Extracted from Hercik et al. (2017) (112).

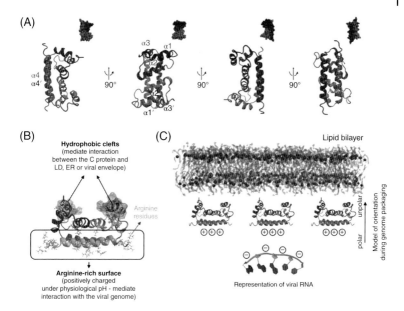

Figure 6.11 Structural aspects of the flavivirus capsid protein. (A) Cartoon of the capsid dimer showing the four alpha helices (α1–α4) in each monomer (PDB ID: 1R6R). Chain A is shown in dark gray and chain B in light gray. Surface representation of each pose is shown on top right. (B) Non-polar and polar regions of the capsid protein. Hydrophobic clefts are exhibited as transparent surfaces. Arginine residues located in the α4-α4′ region are represented by the ball-and-stick models. (C) Orientation model of the capsid protein during virus assembly. *Source:* Extracted from Oliveira et al. (2017) (191).

6.7.2 Membrane (prM/M) Glycoprotein

The glycoprotein precursor of M, or prM, is an approximately 26 kDa glycoprotein while the M protein has a molecular weight of approximately 10 kDa after furin cleavage. prM is translocated into the ER lumen via a signal sequence at the C-terminus of C protein hydrophobic tail. The cleavage of signal peptide is tightly coordinated and dependent on prior cleavage of C by viral NS2B/3 protease (118, 119). The N-terminal region of prM protein consists of one to three N-linked glycosylation sites, and six cysteine residues, which are disulfide, whereas the C-terminal TMD is predicted to form a helical domain that may assist in prM/E heterodimer formation (74, 205–207). Crystal structure analysis of DENV pr peptide has shown that pr domain forms a unique fold consisting of seven β-sheets that are linked by disulfide bridges (208). A major function of prM protein is proposed to assist in the chaperone-mediated folding of the E protein (209, 210). In addition, prM also contributes to the pH-dependent conformational rearrangement of prM/E heterodimer spikes in immature virions (211), preventing premature fusion during virion egress (212, 213).

6.7.3 Envelope (E) Glycoprotein

The E protein (~53 kDa) is the main structural component present on the surface of flavivirus virions. E protein is a type I membrane protein containing 12 conserved cysteines that form disulfide bonds (214). It is a class II fusion protein participating in several key steps during the viral life cycle, including virus binding, entry, and membrane fusion in the susceptible cells.

Crystal structure of ZIKV E protein (residues 1-409) has been solved in prefusion conformation (215), which resembles previously reported flavivirus E protein structures (216–218) and consists of three main domains: a central β-barrel (domain I), an elongated finger-like structure (domain II), and a C-terminal immunoglobulin-like module (domain III) (Figure 6.12 B) (215). Domain I forms an eight-stranded β-barrel and contains the potential N154 glycosylation site (Figure 6.12 B). The fusion peptide is located at the tip of domain II (Figure 6.12 B) and remains covered by the pr peptide or buried in a hydrophobic pocket formed by domains I and III of the partner monomer until triggered to insert into the target cell membrane (1, 217). The C-terminal domain III displays an IgG-like fold, projecting slightly from the virion surface. It is thought to be involved in cellular receptor binding, thus making it a major target of the neutralizing antibodies (219).

Upon exposure to low pH, E protein dimers are dissociated into monomeric subunits, prior to the formation of fusogenic trimers (220–222). Crystal structures of flavivirus E protein at the post-fusion conformation show that E protein folds back onto itself, and in the meantime, brings the N-terminal fusion peptide and its associated cellular membrane into the viral membrane in which the C-terminal transmembrane domain is integrated (216, 223). It has been shown that residues that influence the pH threshold for membrane fusion surround the domain I/II pocket (224). Protonation of conserved histidines at the interface of domain I and domain III was demonstrated to contribute to E domain rearrangements in TBEV recombinant subviral particles (RSPs) (225), though mutagenesis studies of WNV failed to identify histidine residues that controlled the switch (226).

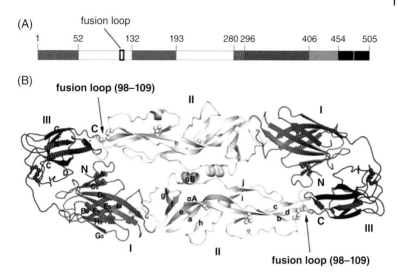

Figure 6.12 Structure of the ZIKV E protein. (A) Domain organization of ZIKV E protein. ZIKV E protein has three distinct domains: β-barrel-shaped domain I, finger-like domain II, and immunoglobulin-like domain III. (B) Dimer structure of ZIKV E proteins. Domain II is responsible for the dimerization of E proteins. The fusion loop is buried by the domains I and II of the other E monomer. *Source:* Extracted from Dai et al. (2016) (215).

6.8 Virus Assembly and Release from Virus-Infected Cells

The assembly process is thought to commence by the association of C protein dimers with genomic RNA. Then virions bud into ER membranes and assemble into an immature form of the virus containing the E-prM glycoprotein complex (1). The immature virus consists of 60 trimeric spikes of E-prM complex; the pr domain of the prM protein protects the fusion loop on the E protein from nonproductive interactions within the cell (227, 228). Crystal structure of immature ZIKV shows a spiky appearance with a diameter of approximately 60 nm containing the trimeric E-prM spike held together at its tip through interactions between the pr domain and fusion loop of one of the E proteins (227).

Following virus assembly, nascent virions are transported through the secretory pathway and released at the cell surface (229). In the low-pH environment of the trans-Golgi complex, immature virions

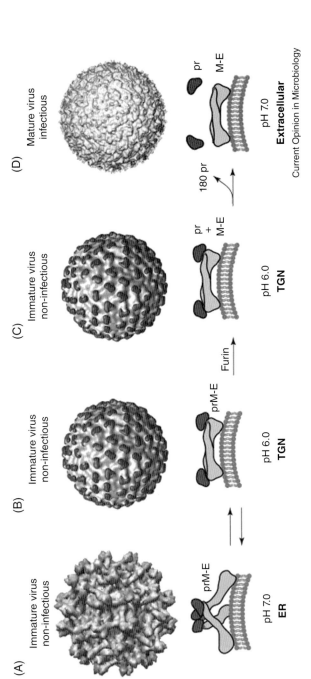

Figure 6.13 Structure of the dengue virion and conformations of the E protein. (A) The cryo-EM reconstruction of the immature virion at neutral pH (239). The E protein forms a heterodimer with prM, and prM-E heterodimers form 60 trimeric spikes that extend away from the surface of the virus. This arrangement of E represents the initial particle that buds into the ER. In this conformation, the pr peptide (dark gray at the top) protects the fusion peptide (intermediate gray at the center) on E (light gray at the bottom). (B) The cryo-EM reconstruction of the immature virion at low pH (240). During its transit through the secretory pathway, the virus encounters low pH in the trans-Golgi network (TGN). The prM-E heterodimers dissociate and form 90 dimers that lie flat against the viral surface. (C) While in the TGN, the prM protein is cleaved by host endoprotease furin into pr peptide and M protein. The cleaved pr peptide maintains its position as a cap on E, while E proteins remain as 90 homodimers lying parallel to the virion surface. M protein lies embedded in the viral membrane beneath the E protein shell. (D) The cryo-EM reconstruction of the mature virion (241). Following furin cleavage, the mature virion is secreted into the extracellular milieu and the pr peptide is released from mature particle.

undergo acid-induced E-prM rearrangement and proteolytic process by host furin protease; the pr domain is cleaved from the prM protein during the maturation (Figure 6.13) (230, 231). Additional maturation steps include glycan modification of prM and E by trimming and terminal addition (108, 232), which is similar to that used for host cell surface glycoproteins. Only the mature virions are considered to be infectious. Because of the varying pr cleavage efficiency, the secreted virion population is a mixture of mature, partially mature, and immature virions (1, 227). It has been shown that immature forms of flaviviruses, including DENV and WNV, become infectious through antibody-dependent entry into the host cells, suggesting a potential role of immature virions in virus infection and spread (233, 234).

References

1 Fields BN, Knipe DM, Howley PM. 2013. Fields virology, 6th ed. Wolters Kluwer Health/Lippincott Williams & Wilkins, Philadelphia.

2 Saiz JC, Vazquez-Calvo A, Blazquez AB, Merino-Ramos T, Escribano-Romero E, Martin-Acebes MA. 2016. Zika Virus: the Latest Newcomer. Front Microbiol 7:496.

3 Kuno G, Chang GJ, Tsuchiya KR, Karabatsos N, Cropp CB. 1998. Phylogeny of the genus Flavivirus. J Virol 72:73–83.

4 Mukhopadhyay S, Kuhn RJ, Rossmann MG. 2005. A structural perspective of the flavivirus life cycle. Nat Rev Microbiol 3:13–22.

5 Kuhn RJ, Zhang W, Rossmann MG, Pletnev SV, Corver J, Lenches E, Jones CT, Mukhopadhyay S, Chipman PR, Strauss EG, Baker TS, Strauss JH. 2002. Structure of dengue virus: implications for flavivirus organization, maturation, and fusion. Cell 108:717–725.

6 Kostyuchenko VA, Lim EX, Zhang S, Fibriansah G, Ng TS, Ooi JS, Shi J, Lok SM. 2016. Structure of the thermally stable Zika virus. Nature 533:425–428.

7 Bell TM, Field EJ, Narang HK. 1971. Zika virus infection of the central nervous system of mice. Arch Gesamte Virusforsch 35:183–193.

8 Zhang X, Ge P, Yu X, Brannan JM, Bi G, Zhang Q, Schein S, Zhou ZH. 2013. Cryo-EM structure of the mature dengue virus at 3.5—A resolution. Nat Struct Mol Biol 20:105–110.

9 Mukhopadhyay S, Kim BS, Chipman PR, Rossmann MG, Kuhn RJ. 2003. Structure of West Nile virus. Science 302:248.

10 Sirohi D, Chen Z, Sun L, Klose T, Pierson TC, Rossmann MG, Kuhn RJ. 2016. The 3.8 A resolution cryo-EM structure of Zika virus. Science **352**:467–470.

11 Fibriansah G, Ng TS, Kostyuchenko VA, Lee J, Lee S, Wang J, Lok SM. 2013. Structural changes in dengue virus when exposed to a temperature of 37 degrees C. J Virol **87**:7585–7592.

12 Zhang X, Sheng J, Plevka P, Kuhn RJ, Diamond MS, Rossmann MG. 2013. Dengue structure differs at the temperatures of its human and mosquito hosts. Proc Natl Acad Sci U S A **110**:6795–6799.

13 Fibriansah G, Ibarra KD, Ng TS, Smith SA, Tan JL, Lim XN, Ooi JS, Kostyuchenko VA, Wang J, de Silva AM, Harris E, Crowe JE, Jr., Lok SM. 2015. DENGUE VIRUS. Cryo-EM structure of an antibody that neutralizes dengue virus type 2 by locking E protein dimers. Science **349**:88–91.

14 Kostyuchenko VA, Chew PL, Ng TS, Lok SM. 2014. Near-atomic resolution cryo-electron microscopic structure of dengue serotype 4 virus. J Virol **88**:477–482.

15 Mansuy JM, Dutertre M, Mengelle C, Fourcade C, Marchou B, Delobel P, Izopet J, Martin-Blondel G. 2016. Zika virus: high infectious viral load in semen, a new sexually transmitted pathogen? Lancet Infect Dis **16**:405.

16 Barzon L, Pacenti M, Berto A, Sinigaglia A, Franchin E, Lavezzo E, Brugnaro P, Palu G. 2016. Isolation of infectious Zika virus from saliva and prolonged viral RNA shedding in a traveller returning from the Dominican Republic to Italy, January 2016. Euro Surveill **21**:30159.

17 Gourinat AC, O'Connor O, Calvez E, Goarant C, Dupont-Rouzeyrol M. 2015. Detection of Zika virus in urine. Emerg Infect Dis **21**:84–86.

18 El Costa H, Gouilly J, Mansuy JM, Chen Q, Levy C, Cartron G, Veas F, Al-Daccak R, Izopet J, Jabrane-Ferrat N. 2016. ZIKA virus reveals broad tissue and cell tropism during the first trimester of pregnancy. Sci Rep **6**:35296.

19 Garcez PP, Loiola EC, Madeiro da Costa R, Higa LM, Trindade P, Delvecchio R, Nascimento JM, Brindeiro R, Tanuri A, Rehen SK. 2016. Zika virus impairs growth in human neurospheres and brain organoids. Science **352**:816–818.

20 Navarro-Sanchez E, Altmeyer R, Amara A, Schwartz O, Fieschi F, Virelizier JL, Arenzana-Seisdedos F, Despres P. 2003. Dendritic-cell-specific ICAM3-grabbing non-integrin is essential for the productive infection of human dendritic cells by mosquito-cell-derived dengue viruses. EMBO Rep **4**:723–728.

21 Hamel R, Dejarnac O, Wichit S, Ekchariyawat P, Neyret A, Luplertlop N, Perera-Lecoin M, Surasombatpattana P, Talignani L, Thomas F, Cao-Lormeau VM, Choumet V, Briant L, Despres P, Amara A, Yssel H, Misse D. 2015. Biology of Zika Virus Infection in Human Skin Cells. J Virol **89**:8880–8896.

22 McGreal EP, Miller JL, Gordon S. 2005. Ligand recognition by antigen-presenting cell C–type lectin receptors. Curr Opin Immunol **17**:18–24.

23 Khoo US, Chan KY, Chan VS, Lin CL. 2008. DC-SIGN and L-SIGN: the SIGNs for infection. J Mol Med (Berl) **86**:861–874.

24 Geijtenbeek TB, Kwon DS, Torensma R, van Vliet SJ, van Duijnhoven GC, Middel J, Cornelissen IL, Nottet HS, KewalRamani VN, Littman DR, Figdor CG, van Kooyk Y. 2000. DC-SIGN, a dendritic cell-specific HIV-1-binding protein that enhances trans-infection of T cells. Cell **100**:587–597.

25 Geijtenbeek TB, Torensma R, van Vliet SJ, van Duijnhoven GC, Adema GJ, van Kooyk Y, Figdor CG. 2000. Identification of DC-SIGN, a novel dendritic cell-specific ICAM-3 receptor that supports primary immune responses. Cell **100**:575–585.

26 Wu SJ, Grouard-Vogel G, Sun W, Mascola JR, Brachtel E, Putvatana R, Louder MK, Filgueira L, Marovich MA, Wong HK, Blauvelt A, Murphy GS, Robb ML, Innes BL, Birx DL, Hayes CG, Frankel SS. 2000. Human skin Langerhans cells are targets of dengue virus infection. Nat Med **6**:816–820.

27 Marovich M, Grouard-Vogel G, Louder M, Eller M, Sun W, Wu SJ, Putvatana R, Murphy G, Tassaneetrithep B, Burgess T, Birx D, Hayes C, Schlesinger-Frankel S, Mascola J. 2001. Human dendritic cells as targets of dengue virus infection. J Investig Dermatol Symp Proc **6**:219–224.

28 Tassaneetrithep B, Burgess TH, Granelli-Piperno A, Trumpfheller C, Finke J, Sun W, Eller MA, Pattanapanyasat K, Sarasombath S, Birx DL, Steinman RM, Schlesinger S, Marovich MA. 2003. DC-SIGN (CD209) mediates dengue virus infection of human dendritic cells. J Exp Med **197**:823–829.

29 Lai C, Lemke G. 1991. An extended family of protein-tyrosine kinase genes differentially expressed in the vertebrate nervous system. Neuron **6**:691–704.

30 Anderson HA, Maylock CA, Williams JA, Paweletz CP, Shu H, Shacter E. 2003. Serum-derived protein S binds to phosphatidylserine and stimulates the phagocytosis of apoptotic cells. Nat Immunol **4**:87–91.

31 Rothlin CV, Ghosh S, Zuniga EI, Oldstone MB, Lemke G. 2007. TAM receptors are pleiotropic inhibitors of the innate immune response. Cell **131**:1124–1136.

32 Rothlin CV, Lemke G. 2010. TAM receptor signaling and autoimmune disease. Curr Opin Immunol **22**:740–746.

33 Nowakowski TJ, Pollen AA, Di Lullo E, Sandoval-Espinosa C, Bershteyn M, Kriegstein AR. 2016. Expression Analysis Highlights AXL as a Candidate Zika Virus Entry Receptor in Neural Stem Cells. Cell Stem Cell **18**:591–596.

34 Liu S, DeLalio LJ, Isakson BE, Wang TT. 2016. AXL-Mediated Productive Infection of Human Endothelial Cells by Zika Virus. Circ Res **119**:1183–1189.

35 Aldo P, You Y, Szigeti K, Horvath TL, Lindenbach B, Mor G. 2016. HSV-2 enhances ZIKV infection of the placenta and induces apoptosis in first-trimester trophoblast cells. Am J Reprod Immunol **76**:348–357.

36 McIntire JJ, Umetsu DT, DeKruyff RH. 2004. TIM-1, a novel allergy and asthma susceptibility gene. Springer Semin Immunopathol **25**:335–348.

37 Meertens L, Carnec X, Lecoin MP, Ramdasi R, Guivel-Benhassine F, Lew E, Lemke G, Schwartz O, Amara A. 2012. The TIM and TAM families of phosphatidylserine receptors mediate dengue virus entry. Cell Host Microbe **12**:544–557.

38 Jemielity S, Wang JJ, Chan YK, Ahmed AA, Li W, Monahan S, Bu X, Farzan M, Freeman GJ, Umetsu DT, Dekruyff RH, Choe H. 2013. TIM-family proteins promote infection of multiple enveloped viruses through virion-associated phosphatidylserine. PLoS Pathog **9**:e1003232.

39 Dick GW, Kitchen SF, Haddow AJ. 1952. Zika virus. I. Isolations and serological specificity. Trans R Soc Trop Med Hyg **46**:509–520.

40 Delvecchio R, Higa LM, Pezzuto P, Valadao AL, Garcez PP, Monteiro FL, Loiola EC, Dias AA, Silva FJ, Aliota MT, Caine EA, Osorio JE, Bellio M, O'Connor DH, Rehen S, de Aguiar RS, Savarino A, Campanati L, Tanuri A. 2016. Chloroquine, an Endocytosis Blocking Agent, Inhibits Zika Virus Infection in Different Cell Models. Viruses **8**.

41 Savidis G, McDougall WM, Meraner P, Perreira JM, Portmann JM, Trincucci G, John SP, Aker AM, Renzette N, Robbins DR, Guo Z, Green S, Kowalik TF, Brass AL. 2016. Identification of Zika Virus and Dengue Virus Dependency Factors using Functional Genomics. Cell Rep **16**:232–246.

42 Kuno G, Chang GJ. 2007. Full-length sequencing and genomic characterization of Bagaza, Kedougou, and Zika viruses. Arch Virol **152**:687–696.

43 Lichinchi G, Zhao BS, Wu Y, Lu Z, Qin Y, He C, Rana TM. 2016. Dynamics of Human and Viral RNA Methylation during Zika Virus Infection. Cell Host Microbe **20**:666–673.

44 Meyer KD, Patil DP, Zhou J, Zinoviev A, Skabkin MA, Elemento O, Pestova TV, Qian SB, Jaffrey SR. 2015. 5′ UTR m(6)A Promotes Cap-Independent Translation. Cell **163**:999–1010.

45 Dominissini D, Moshitch-Moshkovitz S, Schwartz S, Salmon-Divon M, Ungar L, Osenberg S, Cesarkas K, Jacob-Hirsch J, Amariglio N, Kupiec M, Sorek R, Rechavi G. 2012. Topology of the human and mouse m6A RNA methylomes revealed by m6A-seq. Nature **485**:201–206.

46 Schwartz S, Mumbach MR, Jovanovic M, Wang T, Maciag K, Bushkin GG, Mertins P, Ter-Ovanesyan D, Habib N, Cacchiarelli D, Sanjana NE, Freinkman E, Pacold ME, Satija R, Mikkelsen TS, Hacohen N, Zhang F, Carr SA, Lander ES, Regev A. 2014. Perturbation of m6A writers reveals two distinct classes of mRNA methylation at internal and 5′ sites. Cell Rep **8**:284–296.

47 Gokhale NS, McIntyre AB, McFadden MJ, Roder AE, Kennedy EM, Gandara JA, Hopcraft SE, Quicke KM, Vazquez C, Willer J, Ilkayeva OR, Law BA, Holley CL, Garcia-Blanco MA, Evans MJ, Suthar MS, Bradrick SS, Mason CE, Horner SM. 2016. N6-Methyladenosine in Flaviviridae Viral RNA Genomes Regulates Infection. Cell Host Microbe **20**:654–665.

48 Pijlman GP, Funk A, Kondratieva N, Leung J, Torres S, van der Aa L, Liu WJ, Palmenberg AC, Shi PY, Hall RA, Khromykh AA. 2008. A highly structured, nuclease-resistant, noncoding RNA produced by flaviviruses is required for pathogenicity. Cell Host Microbe **4**:579–591.

49 Akiyama BM, Laurence HM, Massey AR, Costantino DA, Xie X, Yang Y, Shi PY, Nix JC, Beckham JD, Kieft JS. 2016. Zika virus produces noncoding RNAs using a multi-pseudoknot structure that confounds a cellular exonuclease. Science **354**:1148–1152.

50 Quicke KM, Bowen JR, Johnson EL, McDonald CE, Ma H, O'Neal JT, Rajakumar A, Wrammert J, Rimawi BH, Pulendran B, Schinazi RF, Chakraborty R, Suthar MS. 2016. Zika Virus Infects Human Placental Macrophages. Cell Host Microbe **20**:83–90.

51 Gokhale NS, Horner SM. 2017. Knotty Zika Virus Blocks Exonuclease to Produce Subgenomic Flaviviral RNAs. Cell Host Microbe **21**:1–2.

52 Kieft JS, Rabe JL, Chapman EG. 2015. New hypotheses derived from the structure of a flaviviral Xrn1-resistant RNA: Conservation, folding, and host adaptation. RNA Biol **12**:1169–1177.

53 Manokaran G, Finol E, Wang C, Gunaratne J, Bahl J, Ong EZ, Tan HC, Sessions OM, Ward AM, Gubler DJ, Harris E, Garcia-Blanco MA, Ooi EE. 2015. Dengue subgenomic RNA binds TRIM25 to inhibit interferon expression for epidemiological fitness. Science **350**:217–221.

54 Schuessler A, Funk A, Lazear HM, Cooper DA, Torres S, Daffis S, Jha BK, Kumagai Y, Takeuchi O, Hertzog P, Silverman R, Akira S, Barton DJ, Diamond MS, Khromykh AA. 2012. West Nile virus noncoding subgenomic RNA contributes to viral evasion of the type I interferon-mediated antiviral response. J Virol **86**:5708–5718.

55 Schnettler E, Sterken MG, Leung JY, Metz SW, Geertsema C, Goldbach RW, Vlak JM, Kohl A, Khromykh AA, Pijlman GP. 2012. Noncoding flavivirus RNA displays RNA interference suppressor activity in insect and Mammalian cells. J Virol **86**:13486–13500.

56 Bidet K, Dadlani D, Garcia-Blanco MA. 2014. G3BP1, G3BP2 and CAPRIN1 are required for translation of interferon stimulated mRNAs and are targeted by a dengue virus non-coding RNA. PLoS Pathog **10**:e1004242.

57 Moon SL, Anderson JR, Kumagai Y, Wilusz CJ, Akira S, Khromykh AA, Wilusz J. 2012. A noncoding RNA produced by arthropod-borne flaviviruses inhibits the cellular exoribonuclease XRN1 and alters host mRNA stability. RNA **18**:2029–2040.

58 Chapman EG, Costantino DA, Rabe JL, Moon SL, Wilusz J, Nix JC, Kieft JS. 2014. The structural basis of pathogenic subgenomic flavivirus RNA (sfRNA) production. Science **344**:307–310.

59 Chapman EG, Moon SL, Wilusz J, Kieft JS. 2014. RNA structures that resist degradation by Xrn1 produce a pathogenic Dengue virus RNA. Elife **3**:e01892.

60 Fleming AM, Ding Y, Alenko A, Burrows CJ. 2016. Zika Virus Genomic RNA Possesses Conserved G-Quadruplexes Characteristic of the Flaviviridae Family. ACS Infect Dis **2**:674–681.

61 Patel DJ, Phan AT, Kuryavyi V. 2007. Human telomere, oncogenic promoter and 5′-UTR G-quadruplexes: diverse higher order DNA and RNA targets for cancer therapeutics. Nucleic Acids Res **35**:7429–7455.

62 Balasubramanian S, Hurley LH, Neidle S. 2011. Targeting G-quadruplexes in gene promoters: a novel anticancer strategy? Nat Rev Drug Discov **10**:261–275.

63 Rhodes D, Lipps HJ. 2015. G-quadruplexes and their regulatory roles in biology. Nucleic Acids Res **43**:8627–8637.

64 Perrone R, Butovskaya E, Daelemans D, Palu G, Pannecouque C, Richter SN. 2014. Anti-HIV-1 activity of the G-quadruplex ligand BRACO-19. J Antimicrob Chemother **69**:3248–3258.

65 Murat P, Zhong J, Lekieffre L, Cowieson NP, Clancy JL, Preiss T, Balasubramanian S, Khanna R, Tellam J. 2014. G-quadruplexes regulate Epstein-Barr virus-encoded nuclear antigen 1 mRNA translation. Nat Chem Biol **10**:358–364.

66 Horsburgh BC, Kollmus H, Hauser H, Coen DM. 1996. Translational recoding induced by G-rich mRNA sequences that form unusual structures. Cell **86**:949–959.

67 Wang SR, Min YQ, Wang JQ, Liu CX, Fu BS, Wu F, Wu LY, Qiao ZX, Song YY, Xu GH, Wu ZG, Huang G, Peng NF, Huang R, Mao WX, Peng S, Chen YQ, Zhu Y, Tian T, Zhang XL, Zhou X. 2016. A highly conserved G-rich consensus sequence in hepatitis C virus core gene represents a new anti-hepatitis C target. Sci Adv **2**:e1501535.

68 Walsh D, Mohr I. 2011. Viral subversion of the host protein synthesis machinery. Nat Rev Microbiol **9**:860–875.

69 Edgil D, Diamond MS, Holden KL, Paranjape SM, Harris E. 2003. Translation efficiency determines differences in cellular infection among dengue virus type 2 strains. Virology **317**:275–290.

70 Edgil D, Polacek C, Harris E. 2006. Dengue virus utilizes a novel strategy for translation initiation when cap-dependent translation is inhibited. J Virol **80**:2976–2986.

71 Roth H, Magg V, Uch F, Mutz P, Klein P, Haneke K, Lohmann V, Bartenschlager R, Fackler OT, Locker N, Stoecklin G, Ruggieri A. 2017. Flavivirus Infection Uncouples Translation Suppression from Cellular Stress Responses. MBio **8**.

72 Daffis S, Szretter KJ, Schriewer J, Li J, Youn S, Errett J, Lin TY, Schneller S, Zust R, Dong H, Thiel V, Sen GC, Fensterl V, Klimstra WB, Pierson TC, Buller RM, Gale M, Jr., Shi PY, Diamond MS. 2010. 2′-O methylation of the viral mRNA cap evades host restriction by IFIT family members. Nature **468**:452–456.

73 Speight G, Coia G, Parker MD, Westaway EG. 1988. Gene mapping and positive identification of the non-structural proteins NS2A, NS2B, NS3, NS4B and NS5 of the flavivirus Kunjin and their cleavage sites. J Gen Virol **69** (Pt 1):23–34.

74 Nowak T, Farber PM, Wengler G. 1989. Analyses of the terminal sequences of West Nile virus structural proteins and of the in vitro translation of these proteins allow the proposal of a complete scheme of the proteolytic cleavages involved in their synthesis. Virology **169**:365–376.

75 Falgout B, Markoff L. 1995. Evidence that flavivirus NS1-NS2A cleavage is mediated by a membrane-bound host protease in the endoplasmic reticulum. J Virol **69**:7232–7243.

76 Shiryaev SA, Ratnikov BI, Chekanov AV, Sikora S, Rozanov DV, Godzik A, Wang J, Smith JW, Huang Z, Lindberg I, Samuel MA, Diamond MS, Strongin AY. 2006. Cleavage targets and the D-arginine-based inhibitors of the West Nile virus NS3 processing proteinase. Biochem J **393**:503–511.

77 Leung D, Schroder K, White H, Fang NX, Stoermer MJ, Abbenante G, Martin JL, Young PR, Fairlie DP. 2001. Activity of recombinant dengue 2 virus NS3 protease in the presence of a truncated NS2B co-factor, small peptide substrates, and inhibitors. J Biol Chem **276**:45762–45771.

78 Bera AK, Kuhn RJ, Smith JL. 2007. Functional characterization of cis and trans activity of the Flavivirus NS2B-NS3 protease. J Biol Chem **282**:12883–12892.

79 Sun G, Larsen CN, Baumgarth N, Klem EB, Scheuermann RH. 2017. Comprehensive Annotation of Mature Peptides and Genotypes for Zika Virus. PLoS One **12**:e0170462.

80 Acosta EG, Kumar A, Bartenschlager R. 2014. Revisiting dengue virus-host cell interaction: new insights into molecular and cellular virology. Adv Virus Res **88**:1–109.

81 Martin-Acebes MA, Saiz JC. 2012. West Nile virus: A re-emerging pathogen revisited. World J Virol **1**:51–70.

82 Xu X, Song H, Qi J, Liu Y, Wang H, Su C, Shi Y, Gao GF. 2016. Contribution of intertwined loop to membrane association revealed by Zika virus full-length NS1 structure. EMBO J **35**:2170–2178.

83 Wang D, Chen C, Liu S, Zhou H, Yang K, Zhao Q, Ji X, Xie W, Wang Z, Mi LZ, Yang H. 2017. A Mutation Identified in Neonatal Microcephaly Destabilizes Zika Virus NS1 Assembly in Vitro. Sci Rep **7**:42580.

84 Song H, Qi J, Haywood J, Shi Y, Gao GF. 2016. Zika virus NS1 structure reveals diversity of electrostatic surfaces among flaviviruses. Nat Struct Mol Biol **23**:456–458.

85 Winkler G, Maxwell SE, Ruemmler C, Stollar V. 1989. Newly synthesized dengue-2 virus nonstructural protein NS1 is a soluble protein but becomes partially hydrophobic and membrane-associated after dimerization. Virology **171**:302–305.

86 Winkler G, Randolph VB, Cleaves GR, Ryan TE, Stollar V. 1988. Evidence that the mature form of the flavivirus nonstructural protein NS1 is a dimer. Virology **162**:187–196.

87 Flamand M, Megret F, Mathieu M, Lepault J, Rey FA, Deubel V. 1999. Dengue virus type 1 nonstructural glycoprotein NS1 is secreted from mammalian cells as a soluble hexamer in a glycosylation-dependent fashion. J Virol **73**:6104–6110.

88 Lindenbach BD, Rice CM. 1999. Genetic interaction of flavivirus nonstructural proteins NS1 and NS4A as a determinant of replicase function. J Virol **73**:4611–4621.

89 Mackenzie JM, Jones MK, Young PR. 1996. Immunolocalization of the dengue virus nonstructural glycoprotein NS1 suggests a role in viral RNA replication. Virology **220**:232–240.

90 Gutsche I, Coulibaly F, Voss JE, Salmon J, d'Alayer J, Ermonval M, Larquet E, Charneau P, Krey T, Megret F, Guittet E, Rey FA, Flamand M. 2011. Secreted dengue virus nonstructural protein NS1 is an atypical barrel-shaped high-density lipoprotein. Proc Natl Acad Sci U S A **108**:8003–8008.

91 Suthar MS, Diamond MS, Gale M, Jr. 2013. West Nile virus infection and immunity. Nat Rev Microbiol **11**:115–128.

92 Youn S, Li T, McCune BT, Edeling MA, Fremont DH, Cristea IM, Diamond MS. 2012. Evidence for a genetic and physical interaction between nonstructural proteins NS1 and NS4B that modulates replication of West Nile virus. J Virol **86**:7360–7371.

93 Young PR, Hilditch PA, Bletchly C, Halloran W. 2000. An antigen capture enzyme-linked immunosorbent assay reveals high levels of the dengue virus protein NS1 in the sera of infected patients. J Clin Microbiol **38**:1053–1057.

94 Chung KM, Nybakken GE, Thompson BS, Engle MJ, Marri A, Fremont DH, Diamond MS. 2006. Antibodies against West Nile Virus nonstructural protein NS1 prevent lethal infection through Fc gamma receptor-dependent and -independent mechanisms. J Virol **80**:1340–1351.

95 Avirutnan P, Fuchs A, Hauhart RE, Somnuke P, Youn S, Diamond MS, Atkinson JP. 2010. Antagonism of the complement component C4 by flavivirus nonstructural protein NS1. J Exp Med **207**:793–806.

96 Avirutnan P, Hauhart RE, Somnuke P, Blom AM, Diamond MS, Atkinson JP. 2011. Binding of flavivirus nonstructural protein NS1 to C4b binding protein modulates complement activation. J Immunol **187**:424–433.

97 Beatty PR, Puerta-Guardo H, Killingbeck SS, Glasner DR, Hopkins K, Harris E. 2015. Dengue virus NS1 triggers endothelial permeability and vascular leak that is prevented by NS1 vaccination. Sci Transl Med 7:304ra141.

98 Modhiran N, Watterson D, Muller DA, Panetta AK, Sester DP, Liu L, Hume DA, Stacey KJ, Young PR. 2015. Dengue virus NS1 protein activates cells via Toll-like receptor 4 and disrupts endothelial cell monolayer integrity. Sci Transl Med 7:304ra142.

99 Scaturro P, Cortese M, Chatel-Chaix L, Fischl W, Bartenschlager R. 2015. Dengue Virus Non-structural Protein 1 Modulates Infectious Particle Production via Interaction with the Structural Proteins. PLoS Pathog **11**:e1005277.

100 Xie X, Gayen S, Kang C, Yuan Z, Shi PY. 2013. Membrane topology and function of dengue virus NS2A protein. J Virol **87**:4609–4622.

101 Mackenzie JM, Khromykh AA, Jones MK, Westaway EG. 1998. Subcellular localization and some biochemical properties of the flavivirus Kunjin nonstructural proteins NS2A and NS4A. Virology **245**:203–215.

102 Kummerer BM, Rice CM. 2002. Mutations in the yellow fever virus nonstructural protein NS2A selectively block production of infectious particles. J Virol **76**:4773–4784.

103 Leung JY, Pijlman GP, Kondratieva N, Hyde J, Mackenzie JM, Khromykh AA. 2008. Role of nonstructural protein NS2A in flavivirus assembly. J Virol **82**:4731–4741.

104 Liu WJ, Chen HB, Wang XJ, Huang H, Khromykh AA. 2004. Analysis of adaptive mutations in Kunjin virus replicon RNA reveals a novel role for the flavivirus nonstructural protein NS2A in inhibition of beta interferon promoter-driven transcription. J Virol **78**:12225–12235.

105 Liu WJ, Wang XJ, Clark DC, Lobigs M, Hall RA, Khromykh AA. 2006. A single amino acid substitution in the West Nile virus nonstructural protein NS2A disables its ability to inhibit alpha/beta interferon induction and attenuates virus virulence in mice. J Virol **80**:2396–2404.

106 Munoz-Jordan JL, Sanchez-Burgos GG, Laurent-Rolle M, Garcia-Sastre A. 2003. Inhibition of interferon signaling by dengue virus. Proc Natl Acad Sci U S A **100**:14333–14338.

107 Tu YC, Yu CY, Liang JJ, Lin E, Liao CL, Lin YL. 2012. Blocking double-stranded RNA-activated protein kinase PKR by Japanese encephalitis virus nonstructural protein 2A. J Virol **86**:10347–10358.

108 Chambers TJ, McCourt DW, Rice CM. 1990. Production of yellow fever virus proteins in infected cells: identification of discrete polyprotein species and analysis of cleavage kinetics using region-specific polyclonal antisera. Virology **177**:159–174.

109 Nestorowicz A, Chambers TJ, Rice CM. 1994. Mutagenesis of the yellow fever virus NS2A/2B cleavage site: effects on proteolytic processing, viral replication, and evidence for alternative processing of the NS2A protein. Virology **199**:114–123.

110 Melian EB, Hinzman E, Nagasaki T, Firth AE, Wills NM, Nouwens AS, Blitvich BJ, Leung J, Funk A, Atkins JF, Hall R, Khromykh AA. 2010. NS1' of flaviviruses in the Japanese encephalitis virus serogroup is a product of ribosomal frameshifting and plays a role in viral neuroinvasiveness. J Virol **84**:1641–1647.

111 Clum S, Ebner KE, Padmanabhan R. 1997. Cotranslational membrane insertion of the serine proteinase precursor NS2B-NS3(Pro) of dengue virus type 2 is required for efficient in vitro processing and is mediated through the hydrophobic regions of NS2B. J Biol Chem **272**:30715–30723.

112 Falgout B, Pethel M, Zhang YM, Lai CJ. 1991. Both nonstructural proteins NS2B and NS3 are required for the proteolytic processing of dengue virus nonstructural proteins. J Virol **65**:2467–2475.

113 Luo D, Xu T, Hunke C, Gruber G, Vasudevan SG, Lescar J. 2008. Crystal structure of the NS3 protease-helicase from dengue virus. J Virol **82**:173–183.

114 Erbel P, Schiering N, D'Arcy A, Renatus M, Kroemer M, Lim SP, Yin Z, Keller TH, Vasudevan SG, Hommel U. 2006. Structural basis for the activation of flaviviral NS3 proteases from dengue and West Nile virus. Nat Struct Mol Biol **13**:372–373.

115 Noble CG, Seh CC, Chao AT, Shi PY. 2012. Ligand-bound structures of the dengue virus protease reveal the active conformation. J Virol **86**:438–446.

116 Robin G, Chappell K, Stoermer MJ, Hu SH, Young PR, Fairlie DP, Martin JL. 2009. Structure of West Nile virus NS3 protease: ligand

stabilization of the catalytic conformation. J Mol Biol **385**:1568–1577.

117 Chambers TJ, Hahn CS, Galler R, Rice CM. 1990. Flavivirus genome organization, expression, and replication. Annu Rev Microbiol **44**:649–688.

118 Amberg SM, Nestorowicz A, McCourt DW, Rice CM. 1994. NS2B-3 proteinase-mediated processing in the yellow fever virus structural region: in vitro and in vivo studies. J Virol **68**:3794–3802.

119 Yamshchikov VF, Compans RW. 1994. Processing of the intracellular form of the west Nile virus capsid protein by the viral NS2B-NS3 protease: an in vitro study. J Virol **68**:5765–5771.

120 Lin C, Amberg SM, Chambers TJ, Rice CM. 1993. Cleavage at a novel site in the NS4A region by the yellow fever virus NS2B-3 proteinase is a prerequisite for processing at the downstream 4A/4B signalase site. J Virol **67**:2327–2335.

121 Zhang Z, Li Y, Loh YR, Phoo WW, Hung AW, Kang C, Luo D. 2016. Crystal structure of unlinked NS2B-NS3 protease from Zika virus. Science **354**:1597–1600.

122 Singleton MR, Dillingham MS, Wigley DB. 2007. Structure and mechanism of helicases and nucleic acid translocases. Annu Rev Biochem **76**:23–50.

123 Jain R, Coloma J, Garcia-Sastre A, Aggarwal AK. 2016. Structure of the NS3 helicase from Zika virus. Nat Struct Mol Biol **23**:752–754.

124 Tian H, Ji X, Yang X, Xie W, Yang K, Chen C, Wu C, Chi H, Mu Z, Wang Z, Yang H. 2016. The crystal structure of Zika virus helicase: basis for antiviral drug design. Protein Cell **7**:450–454.

125 Caruthers JM, McKay DB. 2002. Helicase structure and mechanism. Curr Opin Struct Biol **12**:123–133.

126 Wengler G. 1993. The NS 3 nonstructural protein of flaviviruses contains an RNA triphosphatase activity. Virology **197**:265–273.

127 Luo D, Xu T, Watson RP, Scherer-Becker D, Sampath A, Jahnke W, Yeong SS, Wang CH, Lim SP, Strongin A, Vasudevan SG, Lescar J. 2008. Insights into RNA unwinding and ATP hydrolysis by the flavivirus NS3 protein. EMBO J **27**:3209–3219.

128 Mastrangelo E, Bolognesi M, Milani M. 2012. Flaviviral helicase: insights into the mechanism of action of a motor protein. Biochem Biophys Res Commun **417**:84–87.

129 Tay MY, Saw WG, Zhao Y, Chan KW, Singh D, Chong Y, Forwood JK, Ooi EE, Gruber G, Lescar J, Luo D, Vasudevan SG. 2015. The C-terminal 50 amino acid residues of dengue NS3 protein are

important for NS3-NS5 interaction and viral replication. J Biol Chem **290**:2379–2394.

130 Prikhod'ko GG, Prikhod'ko EA, Pletnev AG, Cohen JI. 2002. Langat flavivirus protease NS3 binds caspase-8 and induces apoptosis. J Virol **76**:5701–5710.

131 Ramanathan MP, Chambers JA, Pankhong P, Chattergoon M, Attatippaholkun W, Dang K, Shah N, Weiner DB. 2006. Host cell killing by the West Nile Virus NS2B-NS3 proteolytic complex: NS3 alone is sufficient to recruit caspase-8-based apoptotic pathway. Virology **345**:56–72.

132 Shafee N, AbuBakar S. 2003. Dengue virus type 2 NS3 protease and NS2B-NS3 protease precursor induce apoptosis. J Gen Virol **84**:2191–2195.

133 Rodriguez-Madoz JR, Belicha-Villanueva A, Bernal-Rubio D, Ashour J, Ayllon J, Fernandez-Sesma A. 2010. Inhibition of the type I interferon response in human dendritic cells by dengue virus infection requires a catalytically active NS2B3 complex. J Virol **84**:9760–9774.

134 Zou J, Xie X, Wang QY, Dong H, Lee MY, Kang C, Yuan Z, Shi PY. 2015. Characterization of dengue virus NS4A and NS4B protein interaction. J Virol **89**:3455–3470.

135 Miller S, Kastner S, Krijnse-Locker J, Buhler S, Bartenschlager R. 2007. The non-structural protein 4A of dengue virus is an integral membrane protein inducing membrane alterations in a 2 K-regulated manner. J Biol Chem **282**:8873–8882.

136 Miller S, Sparacio S, Bartenschlager R. 2006. Subcellular localization and membrane topology of the Dengue virus type 2 Non-structural protein 4B. J Biol Chem **281**:8854–8863.

137 Roosendaal J, Westaway EG, Khromykh A, Mackenzie JM. 2006. Regulated cleavages at the West Nile virus NS4A-2 K-NS4B junctions play a major role in rearranging cytoplasmic membranes and Golgi trafficking of the NS4A protein. J Virol **80**:4623–4632.

138 Teo CS, Chu JJ. 2014. Cellular vimentin regulates construction of dengue virus replication complexes through interaction with NS4A protein. J Virol **88**:1897–1913.

139 Liang Q, Luo Z, Zeng J, Chen W, Foo SS, Lee SA, Ge J, Wang S, Goldman SA, Zlokovic BV, Zhao Z, Jung JU. 2016. Zika Virus NS4A and NS4B Proteins Deregulate Akt-mTOR Signaling in Human Fetal Neural Stem Cells to Inhibit Neurogenesis and Induce Autophagy. Cell Stem Cell **19**:663–671.

140 McLean JE, Wudzinska A, Datan E, Quaglino D, Zakeri Z. 2011. Flavivirus NS4A-induced autophagy protects cells against death and enhances virus replication. J Biol Chem **286**:22147–22159.

141 Shiryaev SA, Chernov AV, Aleshin AE, Shiryaeva TN, Strongin AY. 2009. NS4A regulates the ATPase activity of the NS3 helicase: a novel cofactor role of the non-structural protein NS4A from West Nile virus. J Gen Virol **90**:2081–2085.

142 Umareddy I, Chao A, Sampath A, Gu F, Vasudevan SG. 2006. Dengue virus NS4B interacts with NS3 and dissociates it from single-stranded RNA. J Gen Virol **87**:2605–2614.

143 Zou J, Lee le T, Wang QY, Xie X, Lu S, Yau YH, Yuan Z, Geifman Shochat S, Kang C, Lescar J, Shi PY. 2015. Mapping the Interactions between the NS4B and NS3 proteins of dengue virus. J Virol **89**:3471–3483.

144 Tajima S, Takasaki T, Kurane I. 2011. Restoration of replication-defective dengue type 1 virus bearing mutations in the N-terminal cytoplasmic portion of NS4A by additional mutations in NS4B. Arch Virol **156**:63–69.

145 Ambrose RL, Mackenzie JM. 2011. West Nile virus differentially modulates the unfolded protein response to facilitate replication and immune evasion. J Virol **85**:2723–2732.

146 Munoz-Jordan JL, Laurent-Rolle M, Ashour J, Martinez-Sobrido L, Ashok M, Lipkin WI, Garcia-Sastre A. 2005. Inhibition of alpha/beta interferon signaling by the NS4B protein of flaviviruses. J Virol **79**:8004–8013.

147 Saeedi BJ, Geiss BJ. 2013. Regulation of flavivirus RNA synthesis and capping. Wiley Interdiscip Rev RNA **4**:723–735.

148 Ray D, Shah A, Tilgner M, Guo Y, Zhao Y, Dong H, Deas TS, Zhou Y, Li H, Shi PY. 2006. West Nile virus 5′-cap structure is formed by sequential guanine N-7 and ribose 2′-O methylations by nonstructural protein 5. J Virol **80**:8362–8370.

149 Issur M, Geiss BJ, Bougie I, Picard-Jean F, Despins S, Mayette J, Hobdey SE, Bisaillon M. 2009. The flavivirus NS5 protein is a true RNA guanylyltransferase that catalyzes a two-step reaction to form the RNA cap structure. RNA **15**:2340–2350.

150 Ackermann M, Padmanabhan R. 2001. De novo synthesis of RNA by the dengue virus RNA-dependent RNA polymerase exhibits temperature dependence at the initiation but not elongation phase. J Biol Chem **276**:39926–39937.

151 Kao CC, Singh P, Ecker DJ. 2001. De novo initiation of viral RNA-dependent RNA synthesis. Virology **287**:251–260.

152 van Dijk AA, Makeyev EV, Bamford DH. 2004. Initiation of viral RNA-dependent RNA polymerization. J Gen Virol **85**:1077–1093.

153 Zhao B, Yi G, Du F, Chuang YC, Vaughan RC, Sankaran B, Kao CC, Li P. 2017. Structure and function of the Zika virus full-length NS5 protein. Nat Commun **8**:14762.

154 Wang B, Tan XF, Thurmond S, Zhang ZM, Lin A, Hai R, Song J. 2017. The structure of Zika virus NS5 reveals a conserved domain conformation. Nat Commun **8**:14763.

155 Upadhyay AK, Cyr M, Longenecker K, Tripathi R, Sun C, Kempf DJ. 2017. Crystal structure of full-length Zika virus NS5 protein reveals a conformation similar to Japanese encephalitis virus NS5. Acta Crystallogr F Struct Biol Commun **73**:116–122.

156 Duan W, Song H, Wang H, Chai Y, Su C, Qi J, Shi Y, Gao GF. 2017. The crystal structure of Zika virus NS5 reveals conserved drug targets. EMBO J doi:10.15252/embj.201696241.

157 Zhang C, Feng T, Cheng J, Li Y, Yin X, Zeng W, Jin X, Guo F, Jin T. 2016. Structure of the NS5 methyltransferase from Zika virus and implications in inhibitor design. Biochem Biophys Res Commun doi:10.1016/j.bbrc.2016.11.098.

158 Coloma J, Jain R, Rajashankar KR, Garcia-Sastre A, Aggarwal AK. 2016. Structures of NS5 Methyltransferase from Zika Virus. Cell Rep **16**:3097–3102.

159 Ferrer-Orta C, Arias A, Escarmis C, Verdaguer N. 2006. A comparison of viral RNA-dependent RNA polymerases. Curr Opin Struct Biol **16**:27–34.

160 Godoy AS, Lima GM, Oliveira KI, Torres NU, Maluf FV, Guido RV, Oliva G. 2017. Crystal structure of Zika virus NS5 RNA-dependent RNA polymerase. Nat Commun **8**:14764.

161 Dong H, Chang DC, Xie X, Toh YX, Chung KY, Zou G, Lescar J, Lim SP, Shi PY. 2010. Biochemical and genetic characterization of dengue virus methyltransferase. Virology **405**:568–578.

162 Kroschewski H, Lim SP, Butcher RE, Yap TL, Lescar J, Wright PJ, Vasudevan SG, Davidson AD. 2008. Mutagenesis of the dengue virus type 2 NS5 methyltransferase domain. J Biol Chem **283**:19410–19421.

163 Zhou Y, Ray D, Zhao Y, Dong H, Ren S, Li Z, Guo Y, Bernard KA, Shi PY, Li H. 2007. Structure and function of flavivirus NS5 methyltransferase. J Virol **81**:3891–3903.

164 Bhattacharya D, Ansari IH, Striker R. 2009. The flaviviral methyltransferase is a substrate of Casein Kinase 1. Virus Res **141**:101–104.

165 Bhattacharya D, Hoover S, Falk SP, Weisblum B, Vestling M, Striker R. 2008. Phosphorylation of yellow fever virus NS5 alters methyltransferase activity. Virology **380**:276–284.

166 Johansson M, Brooks AJ, Jans DA, Vasudevan SG. 2001. A small region of the dengue virus-encoded RNA-dependent RNA polymerase, NS5, confers interaction with both the nuclear transport receptor importin-beta and the viral helicase, NS3. J Gen Virol **82**:735–745.

167 Kapoor M, Zhang L, Ramachandra M, Kusukawa J, Ebner KE, Padmanabhan R. 1995. Association between NS3 and NS5 proteins of dengue virus type 2 in the putative RNA replicase is linked to differential phosphorylation of NS5. J Biol Chem **270**:19100–19106.

168 Cui T, Sugrue RJ, Xu Q, Lee AK, Chan YC, Fu J. 1998. Recombinant dengue virus type 1 NS3 protein exhibits specific viral RNA binding and NTPase activity regulated by the NS5 protein. Virology **246**:409–417.

169 Yon C, Teramoto T, Mueller N, Phelan J, Ganesh VK, Murthy KH, Padmanabhan R. 2005. Modulation of the nucleoside triphosphatase/RNA helicase and 5′-RNA triphosphatase activities of Dengue virus type 2 nonstructural protein 3 (NS3) by interaction with NS5, the RNA-dependent RNA polymerase. J Biol Chem **280**:27412–27419.

170 Mackenzie JM, Kenney MT, Westaway EG. 2007. West Nile virus strain Kunjin NS5 polymerase is a phosphoprotein localized at the cytoplasmic site of viral RNA synthesis. J Gen Virol **88**:1163–1168.

171 Welsch S, Miller S, Romero-Brey I, Merz A, Bleck CK, Walther P, Fuller SD, Antony C, Krijnse-Locker J, Bartenschlager R. 2009. Composition and three-dimensional architecture of the dengue virus replication and assembly sites. Cell Host Microbe **5**:365–375.

172 Davidson AD. 2009. Chapter 2. New insights into flavivirus nonstructural protein 5. Adv Virus Res **74**:41–101.

173 Best SM, Morris KL, Shannon JG, Robertson SJ, Mitzel DN, Park GS, Boer E, Wolfinbarger JB, Bloom ME. 2005. Inhibition of interferon-stimulated JAK-STAT signaling by a tick-borne flavivirus and identification of NS5 as an interferon antagonist. J Virol **79**:12828–12839.

174 Grant A, Ponia SS, Tripathi S, Balasubramaniam V, Miorin L, Sourisseau M, Schwarz MC, Sanchez-Seco MP, Evans MJ, Best SM, Garcia-Sastre A. 2016. Zika Virus Targets Human STAT2 to Inhibit Type I Interferon Signaling. Cell Host Microbe **19**:882–890.

175 Laurent-Rolle M, Morrison J, Rajsbaum R, Macleod JM, Pisanelli G, Pham A, Ayllon J, Miorin L, Martinez-Romero C, tenOever BR, Garcia-Sastre A. 2014. The interferon signaling antagonist function of yellow fever virus NS5 protein is activated by type I interferon. Cell Host Microbe **16**:314–327.

176 Mazzon M, Jones M, Davidson A, Chain B, Jacobs M. 2009. Dengue virus NS5 inhibits interferon-alpha signaling by blocking signal transducer and activator of transcription 2 phosphorylation. J Infect Dis **200**:1261–1270.

177 Morrison J, Laurent-Rolle M, Maestre AM, Rajsbaum R, Pisanelli G, Simon V, Mulder LC, Fernandez-Sesma A, Garcia-Sastre A. 2013. Dengue virus co-opts UBR4 to degrade STAT2 and antagonize type I interferon signaling. PLoS Pathog **9**:e1003265.

178 Paul D, Bartenschlager R. 2013. Architecture and biogenesis of plus-strand RNA virus replication factories. World J Virol **2**:32–48.

179 Bily T, Palus M, Eyer L, Elsterova J, Vancova M, Ruzek D. 2015. Electron Tomography Analysis of Tick-Borne Encephalitis Virus Infection in Human Neurons. Sci Rep **5**:10745.

180 Gillespie LK, Hoenen A, Morgan G, Mackenzie JM. 2010. The endoplasmic reticulum provides the membrane platform for biogenesis of the flavivirus replication complex. J Virol **84**:10438–10447.

181 Junjhon J, Pennington JG, Edwards TJ, Perera R, Lanman J, Kuhn RJ. 2014. Ultrastructural characterization and three-dimensional architecture of replication sites in dengue virus-infected mosquito cells. J Virol **88**:4687–4697.

182 Miorin L, Romero-Brey I, Maiuri P, Hoppe S, Krijnse-Locker J, Bartenschlager R, Marcello A. 2013. Three-dimensional architecture of tick-borne encephalitis virus replication sites and trafficking of the replicated RNA. J Virol **87**:6469–6481.

183 Offerdahl DK, Dorward DW, Hansen BT, Bloom ME. 2017. Cytoarchitecture of Zika virus infection in human neuroblastoma and Aedes albopictus cell lines. Virology **501**:54–62.

184 Cortese M, Goellner S, Acosta EG, Neufeldt CJ, Oleksiuk O, Lampe M, Haselmann U, Funaya C, Schieber N, Ronchi P, Schorb M, Pruunsild P, Schwab Y, Chatel-Chaix L, Ruggieri A, Bartenschlager R. 2017. Ultrastructural Characterization of Zika Virus Replication Factories. Cell Rep **18**:2113–2123.

185 Lindenbach BD, Rice CM. 1997. trans-Complementation of yellow fever virus NS1 reveals a role in early RNA replication. J Virol **71**:9608–9617.

186 Cleaves GR, Ryan TE, Schlesinger RW. 1981. Identification and characterization of type 2 dengue virus replicative intermediate and replicative form RNAs. Virology **111**:73–83.

187 Muylaert IR, Chambers TJ, Galler R, Rice CM. 1996. Mutagenesis of the N-linked glycosylation sites of the yellow fever virus NS1 protein: effects on virus replication and mouse neurovirulence. Virology **222**:159–168.

188 Chu PW, Westaway EG. 1985. Replication strategy of Kunjin virus: evidence for recycling role of replicative form RNA as template in semiconservative and asymmetric replication. Virology **140**:68–79.

189 Jones CT, Ma L, Burgner JW, Groesch TD, Post CB, Kuhn RJ. 2003. Flavivirus capsid is a dimeric alpha-helical protein. J Virol **77**:7143–7149.

190 Wang SH, Syu WJ, Hu ST. 2004. Identification of the homotypic interaction domain of the core protein of dengue virus type 2. J Gen Virol **85**:2307–2314.

191 Oliveira ER, Mohana-Borges R, de Alencastro RB, Horta BA. 2017. The flavivirus capsid protein: Structure, function and perspectives towards drug design. Virus Res **227**:115–123.

192 Markoff L, Falgout B, Chang A. 1997. A conserved internal hydrophobic domain mediates the stable membrane integration of the dengue virus capsid protein. Virology **233**:105–117.

193 Ma L, Jones CT, Groesch TD, Kuhn RJ, Post CB. 2004. Solution structure of dengue virus capsid protein reveals another fold. Proc Natl Acad Sci U S A **101**:3414–3419.

194 Patkar CG, Jones CT, Chang YH, Warrier R, Kuhn RJ. 2007. Functional requirements of the yellow fever virus capsid protein. J Virol **81**:6471–6481.

195 Kofler RM, Heinz FX, Mandl CW. 2002. Capsid protein C of tick-borne encephalitis virus tolerates large internal deletions and is a favorable target for attenuation of virulence. J Virol **76**:3534–3543.

196 Tsuda Y, Mori Y, Abe T, Yamashita T, Okamoto T, Ichimura T, Moriishi K, Matsuura Y. 2006. Nucleolar protein B23 interacts with Japanese encephalitis virus core protein and participates in viral replication. Microbiol Immunol **50**:225–234.

197 Mori Y, Okabayashi T, Yamashita T, Zhao Z, Wakita T, Yasui K, Hasebe F, Tadano M, Konishi E, Moriishi K, Matsuura Y. 2005. Nuclear localization of Japanese encephalitis virus core protein enhances viral replication. J Virol **79**:3448–3458.

198 Wang SH, Syu WJ, Huang KJ, Lei HY, Yao CW, King CC, Hu ST. 2002. Intracellular localization and determination of a nuclear localization signal of the core protein of dengue virus. J Gen Virol **83**:3093–3102.

199 Colpitts TM, Barthel S, Wang P, Fikrig E. 2011. Dengue virus capsid protein binds core histones and inhibits nucleosome formation in human liver cells. PLoS One **6**:e24365.

200 Chang CJ, Luh HW, Wang SH, Lin HJ, Lee SC, Hu ST. 2001. The heterogeneous nuclear ribonucleoprotein K (hnRNP K) interacts with dengue virus core protein. DNA Cell Biol **20**:569–577.

201 Limjindaporn T, Netsawang J, Noisakran S, Thiemmeca S, Wongwiwat W, Sudsaward S, Avirutnan P, Puttikhunt C, Kasinrerk W, Sriburi R, Sittisombut N, Yenchitsomanus PT, Malasit P. 2007. Sensitization to Fas-mediated apoptosis by dengue virus capsid protein. Biochem Biophys Res Commun **362**:334–339.

202 Netsawang J, Noisakran S, Puttikhunt C, Kasinrerk W, Wongwiwat W, Malasit P, Yenchitsomanus PT, Limjindaporn T. 2010. Nuclear localization of dengue virus capsid protein is required for DAXX interaction and apoptosis. Virus Res **147**:275–283.

203 Yang JS, Ramanathan MP, Muthumani K, Choo AY, Jin SH, Yu QC, Hwang DS, Choo DK, Lee MD, Dang K, Tang W, Kim JJ, Weiner DB. 2002. Induction of inflammation by West Nile virus capsid through the caspase-9 apoptotic pathway. Emerg Infect Dis **8**:1379–1384.

204 Freire JM, Veiga AS, de la Torre BG, Santos NC, Andreu D, Da Poian AT, Castanho MA. 2013. Peptides as models for the structure and function of viral capsid proteins: Insights on dengue virus capsid. Biopolymers **100**:325–336.

205 Hsieh SC, Zou G, Tsai WY, Qing M, Chang GJ, Shi PY, Wang WK. 2011. The C-terminal helical domain of dengue virus precursor membrane protein is involved in virus assembly and entry. Virology **410**:170–180.

206 Peng JG, Wu SC. 2014. Glutamic acid at residue 125 of the prM helix domain interacts with positively charged amino acids in E protein domain II for Japanese encephalitis virus-like-particle production. J Virol **88**:8386–8396.

207 Zhang Q, Hunke C, Yau YH, Seow V, Lee S, Tanner LB, Guan XL, Wenk MR, Fibriansah G, Chew PL, Kukkaro P, Biukovic G, Shi PY, Shochat SG, Gruber G, Lok SM. 2012. The stem region of premembrane protein plays an important role in the virus surface protein rearrangement during dengue maturation. J Biol Chem **287**:40525–40534.

208 Li L, Lok SM, Yu IM, Zhang Y, Kuhn RJ, Chen J, Rossmann MG. 2008. The flavivirus precursor membrane-envelope protein complex: structure and maturation. Science **319**:1830–1834.

209 Konishi E, Mason PW. 1993. Proper maturation of the Japanese encephalitis virus envelope glycoprotein requires cosynthesis with the premembrane protein. J Virol **67**:1672–1675.

210 Lorenz IC, Allison SL, Heinz FX, Helenius A. 2002. Folding and dimerization of tick-borne encephalitis virus envelope proteins prM and E in the endoplasmic reticulum. J Virol **76**:5480–5491.

211 Zhang Y, Kaufmann B, Chipman PR, Kuhn RJ, Rossmann MG. 2007. Structure of immature West Nile virus. J Virol **81**:6141–6145.

212 Guirakhoo F, Bolin RA, Roehrig JT. 1992. The Murray Valley encephalitis virus prM protein confers acid resistance to virus particles and alters the expression of epitopes within the R2 domain of E glycoprotein. Virology **191**:921–931.

213 Heinz FX, Stiasny K, Puschner-Auer G, Holzmann H, Allison SL, Mandl CW, Kunz C. 1994. Structural changes and functional control of the tick-borne encephalitis virus glycoprotein E by the heterodimeric association with protein prM. Virology **198**:109–117.

214 Nowak T, Wengler G. 1987. Analysis of disulfides present in the membrane proteins of the West Nile flavivirus. Virology **156**:127–137.

215 Dai L, Song J, Lu X, Deng YQ, Musyoki AM, Cheng H, Zhang Y, Yuan Y, Song H, Haywood J, Xiao H, Yan J, Shi Y, Qin CF, Qi J, Gao GF. 2016. Structures of the Zika Virus Envelope Protein and Its Complex with a Flavivirus Broadly Protective Antibody. Cell Host Microbe **19**:696–704.

216 Modis Y, Ogata S, Clements D, Harrison SC. 2004. Structure of the dengue virus envelope protein after membrane fusion. Nature **427**:313–319.

217 Rey FA, Heinz FX, Mandl C, Kunz C, Harrison SC. 1995. The envelope glycoprotein from tick-borne encephalitis virus at 2 A resolution. Nature **375**:291–298.

218 Luca VC, AbiMansour J, Nelson CA, Fremont DH. 2012. Crystal structure of the Japanese encephalitis virus envelope protein. J Virol **86**:2337–2346.

219 Chu JJ, Rajamanonmani R, Li J, Bhuvanakantham R, Lescar J, Ng ML. 2005. Inhibition of West Nile virus entry by using a

recombinant domain III from the envelope glycoprotein. J Gen Virol **86**:405–412.

220 Allison SL, Schalich J, Stiasny K, Mandl CW, Kunz C, Heinz FX. 1995. Oligomeric rearrangement of tick-borne encephalitis virus envelope proteins induced by an acidic pH. J Virol **69**:695–700.

221 Stiasny K, Allison SL, Marchler-Bauer A, Kunz C, Heinz FX. 1996. Structural requirements for low-pH-induced rearrangements in the envelope glycoprotein of tick-borne encephalitis virus. J Virol **70**:8142–8147.

222 Stiasny K, Bressanelli S, Lepault J, Rey FA, Heinz FX. 2004. Characterization of a membrane-associated trimeric low-pH-induced Form of the class II viral fusion protein E from tick-borne encephalitis virus and its crystallization. J Virol **78**:3178–3183.

223 Bressanelli S, Stiasny K, Allison SL, Stura EA, Duquerroy S, Lescar J, Heinz FX, Rey FA. 2004. Structure of a flavivirus envelope glycoprotein in its low-pH-induced membrane fusion conformation. EMBO J **23**:728–738.

224 Modis Y, Ogata S, Clements D, Harrison SC. 2003. A ligand-binding pocket in the dengue virus envelope glycoprotein. Proc Natl Acad Sci U S A **100**:6986–6991.

225 Fritz R, Stiasny K, Heinz FX. 2008. Identification of specific histidines as pH sensors in flavivirus membrane fusion. J Cell Biol **183**:353–361.

226 Nelson S, Poddar S, Lin TY, Pierson TC. 2009. Protonation of individual histidine residues is not required for the pH-dependent entry of west nile virus: evaluation of the "histidine switch" hypothesis. J Virol **83**:12631–12635.

227 Prasad VM, Miller AS, Klose T, Sirohi D, Buda G, Jiang W, Kuhn RJ, Rossmann MG. 2017. Structure of the immature Zika virus at 9 A resolution. Nat Struct Mol Biol **24**:184–186.

228 Zhang Y, Corver J, Chipman PR, Zhang W, Pletnev SV, Sedlak D, Baker TS, Strauss JH, Kuhn RJ, Rossmann MG. 2003. Structures of immature flavivirus particles. EMBO J **22**:2604–2613.

229 Mackenzie JM, Westaway EG. 2001. Assembly and maturation of the flavivirus Kunjin virus appear to occur in the rough endoplasmic reticulum and along the secretory pathway, respectively. J Virol **75**:10787–10799.

230 Stadler K, Allison SL, Schalich J, Heinz FX. 1997. Proteolytic activation of tick-borne encephalitis virus by furin. J Virol **71**:8475–8481.

231 Perera R, Kuhn RJ. 2008. Structural proteomics of dengue virus. Curr Opin Microbiol **11**:369–377.

232 Courageot MP, Frenkiel MP, Dos Santos CD, Deubel V, Despres P. 2000. Alpha-glucosidase inhibitors reduce dengue virus production by affecting the initial steps of virion morphogenesis in the endoplasmic reticulum. J Virol **74**:564–572.

233 Rodenhuis-Zybert IA, van der Schaar HM, da Silva Voorham JM, van der Ende-Metselaar H, Lei HY, Wilschut J, Smit JM. 2010. Immature dengue virus: a veiled pathogen? PLoS Pathog **6**:e1000718.

234 Colpitts TM, Rodenhuis-Zybert I, Moesker B, Wang P, Fikrig E, Smit JM. 2011. prM-antibody renders immature West Nile virus infectious in vivo. J Gen Virol **92**:2281–2285.

235 Dereeper A, Guignon V, Blanc G, Audic S, Buffet S, Chevenet F, Dufayard JF, Guindon S, Lefort V, Lescot M, Claverie JM, Gascuel O. 2008. Phylogeny.fr: robust phylogenetic analysis for the non-specialist. Nucleic Acids Res **36**:W465–469.

236 Akey DL, Brown WC, Dutta S, Konwerski J, Jose J, Jurkiw TJ, DelProposto J, Ogata CM, Skiniotis G, Kuhn RJ, Smith JL. 2014. Flavivirus NS1 structures reveal surfaces for associations with membranes and the immune system. Science **343**:881–885.

237 Stuart DI, Levine M, Muirhead H, Stammers DK. 1979. Crystal structure of cat muscle pyruvate kinase at a resolution of 2.6 A. J Mol Biol **134**:109–142.

238 Felsenstein J. 1997. An alternating least squares approach to inferring phylogenies from pairwise distances. Syst Biol **46**:101–111.

239 Zhang Y, Zhang W, Ogata S, Clements D, Strauss JH, Baker TS, Kuhn RJ, Rossmann MG. 2004. Conformational changes of the flavivirus E glycoprotein. Structure **12**:1607–1618.

240 Yu IM, Zhang W, Holdaway HA, Li L, Kostyuchenko VA, Chipman PR, Kuhn RJ, Rossmann MG, Chen J. 2008. Structure of the immature dengue virus at low pH primes proteolytic maturation. Science **319**:1834–1837.

241 Zhang W, Chipman PR, Corver J, Johnson PR, Zhang Y, Mukhopadhyay S, Baker TS, Strauss JH, Rossmann MG, Kuhn RJ. 2003. Visualization of membrane protein domains by cryo-electron microscopy of dengue virus. Nat Struct Biol **10**:907–912.

7

Zika Virus (ZIKV) Strains and Lineages

Since the Zika virus (ZIKV) was discovered in 1947 (1, 2), many strains have been isolated in Africa (3, 4), and more recently worldwide (5–8). Clinical differences among the outbreaks between 1952 and 2015 have generated many questions on how possible modifications in the ZIKV RNA could have resulted in different outcomes after ZIKV infection. Before the French Polynesia outbreak in 2013 (9), there was no report of important neurological problems associated with ZIKV infection in humans (10, 11). No clinical difference among the infected patients from different continents was found, and the symptoms were usually restricted to cutaneous rash, slight fever, arthralgia, and, in some cases, muscle pains and conjunctivitis (11–13).

7.1   East and West African Lineage

ZIKV infection reported in humans from East (Uganda) and West (Nigeria) of Africa were associated with different symptoms, which implied that these strains might not be the same (12, 14, 15). Later, it was clarified that the first two studies from the West referred to an infection by a Spondweni virus (SPOV) and not ZIKV (16). No major consequence was detected in the population naturally exposed to the virus at that time (3). Although some sequencing differences have been reported, the East and West African strains belong to the same cluster, named simply as the "African" from now on (17).

Zika Virus and Diseases: From Molecular Biology to Epidemiology, First Edition.
Suzane Ramos da Silva, Fan Cheng and Shou-Jiang Gao.
© 2018 John Wiley & Sons, Inc. Published 2018 by John Wiley & Sons, Inc.

7.2 Africa vs. Asian/American Lineage

ZIKV isolated from four patients during the Yap Island 2007 outbreak were combined and sequenced. The results were aligned, based on the sequences of NS5 protein, with other viruses such as Dengue virus (DENV) and SPOV, which are closely related to ZIKV, as well as other African ZIKV isolates. All ZIKV isolated were clustered in the same clade, although three branches could be identified as: West African (from Senegal isolated in 1984), MR766 (which was the first ZIKV isolated, from Uganda in 1947) (2), and the Yap Island strain (from 2007)(17). For future references, these branches were denoted as African (West and East) and Asian lineages, respectively, (18).

One of the most important nucleotide modifications observed in flavivirus is at the 154 Asn-X-Ser/Thr glycosylation motif in the envelope protein. This site was conserved in the Yap Island and Senegal strains but was absent in MR766 (17). One important fact is the time associated with the isolation of these ZIKV strains, which spanned almost 70 years. Based on studies with other flaviviruses such as West-Nile virus (WNV), modifications in the amino acid 154 of the envelope protein are associated with the invasiveness of the viruses. Prolonged passage of the virus might be associated with glycosylation and has been reported to cause a decrease or increase of neuro-invasion, indicating that other sites and modifications might also be associated with the pathogenesis (19–21). As previously reported, ZIKV MR766 strain was passaged hundreds of time (22).

Other studies were carried out to identify important alterations that characterized African and Asian lineages. A study by Haddow et al. (2012) compared the sequences among African (Senegal/ArD 41519 and Nigeria/IbH 30656) and Asian strains (Malaysia/P6-740, Micronesia/EC Yap, and Cambodia/FS13025) (Figure 7.1). A low-passage MR766 strain was also included. Interestingly, deletion around the N-linked glycosylation site was observed in the low-passage MR766 and IbH30656 (Figure 7.2) (5). Sequence analysis of isolates from Senegal, Uganda, and Cambodia (FSS13025) identified a deletion of 12 nucleotides in MR766. The presence of this deletion, which was absent in FSS13025, depended on the passage number of the virus, (23). Further analyses of isolates from other outbreaks have added important information about their sequences. Alignment, based on NS5 protein, of the first ZIKV isolated from French Polynesia (H/PF/2013) indicated that it belonged to the Asian lineage. This

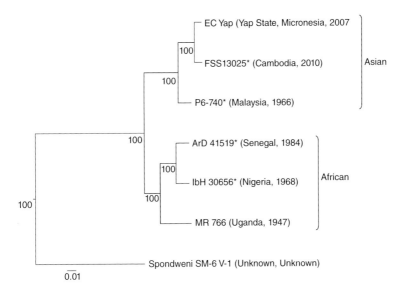

Figure 7.1 Nucleotide and amino acid alignments for different strains of ZIKV. Neighbor-joining phylogeny generated from open reading frame nucleotide sequences of ZIKV strains. The tree was rooted with Spondweni virus (DQ859064). The scale at the bottom of the tree represents genetic distance in nucleotide substitutions per site. Numbers at the nodes represent percent bootstrap support values based on 1,000 replicates. Isolates are represented according to strain name, country of origin, and year of isolation. The lineage of each virus is indicated to the right of the tree. *Source:* *Strains sequenced in Haddow et al. (2012) study. Figure and legend extracted from Haddow et al. (2012) (5).

```
                          450       460       470       480       490
                          |....|....|....|....|....|....|....|....|....|
MR 766      (Uganda, 1947)  GATGATT-----------GGATATGAAACTGACGAAGATA
ArD 41519 (Senegal, 1984)  GATGATTGTGAATGACACAGGACATGAAACTGACGAAAACA
IbH 30656 (Nigeria, 1968)  GATGATTGTGAATGAC-----------------GAAAACA
EC Yap (Micronesia, 2007)  GATGATCGTTAATGACACAGGACATGAAACTGATGAGAATA
P6-740    (Malaysia, 1966)  GATGATTGTTAATGACANAGGACATGAAACTGATGAGAATA
FSS13025  (Cambodia, 2010)  GATGATCGTTAATGATACAGGACATGAAACTGATGAGAATA

                          140       150       160       170       180
                          |....|....|....|....|....|....|....|....|....|
MR 766      (Uganda, 1947)  MLSVHGSQHSGMI----GYETDEDRAKVEVTPNSPRAEATL
ArD 41519 (Senegal, 1984)  MLSVHGSQHSGMIVNDTGHETDENRAKVEVTPNSPRAEATL
IbH 30656 (Nigeria, 1968)  MLSVHGSQHSGMIVND------ENRAKVEVTPNSPRAEATL
EC Yap (Micronesia, 2007)  MLSVHGSQHSGMIVNDTGHETDENRAKVEITPNSPRAEATL
P6-740    (Malaysia, 1966)  MLSVHGSQHSGMIVNDXGHETDENRAKVEITPNSPRAEATL
FSS13025  (Cambodia, 2010)  MLSVHGSQHSGMIVNDTGHETDENRAKVEITPNSPRAEATL
```

Figure 7.2 Zika virus phylogeny. Nucleotide (top) and amino acid (bottom) sequences of the envelope protein/gene of Zika virus strains showing deletions in the potential glycosylation sites of the MR 766 (Uganda, 1947) and the IbH 30656 (Nigeria, 1968) strains. Deletions are indicated by dashes. The "N" at position 467 of the P6-740 strain (Malaysia, 1966) represents an equally weighted double population of the nucleotides "C" and "T." This translates to an "X" at position 165 of the amino acid alignment. *Source:* Figure and legend extracted from Haddow et al. (2012) (5).

strain had a 99.9% similarity to those isolated in Cambodia in 2010 (6) and Yap Island in 2007 (7, 17). ZIKV from Thailand isolated in 2014 (Zika virus/*Homo sapiens*-tc/THA/2014/SV0127-14) and Philippines isolated in 2012 (Zika virus/*H. sapiens*-tc/PHL/2012/CPC-0740) were determined to be of Asian lineage (24). Besides modifications along the ZIKV genome, a structural change of SLI at the 3′UTR was also described in Asian strains (25).

Important differences in the infectivity of iPSCs between African and Asian ZIKV strains were demonstrated with ArB41644 and H/PF/2013 strains, respectively. Infection rate and virus production were higher with the African strain, and the infected cells presented cell cycle impairment and upregulation of Retinoic Acid Inducible Gene 1 Protein (RIG-I), Melanoma Differentiation-Associated protein 5 (MDA5) and Toll-like receptor 3 (TLR3). Both strains had similar passage number, hence the differences observed among these strains were likely biological but not a result of virus handling (26). Other *in vitro* experiments with human primary astrocytes indicated that both African (HD 78788) and Asian (H/PF/2013) strains were able to induce anti-viral response with induction of interferon (IFN)-I, but their kinetics were different. At time-point as early as 6 hours post-infection (h.p.i.), Asian strain was able to induce innate immune response, while this was observed only after 24 h.p.i. for the African strain. Up-regulation of TLR3 was observed with both strains, but African strain also induced overexpression of TLR7 and TLR9. No difference was observed in the expression of viral genes at 48 h.p.i. (27).

ZIKV strains from Asian lineage circulated in South America in late 2015 (8, 28–32). Although strains circulating in South Pacific and the Americas belong to the Asian lineage, the diseases associated with these strains are different (11, 13). Despite the recent outbreak in the Americas, modifications have already been reported among the viruses isolated from different Central and South America countries, indicating a dynamic evolution nature of the Asian strains compared to the African strains (33). ZIKV isolated from Martinique was closely related to French Polynesia, while those from Brazil had a few modifications mostly in the membrane gene (34).

In Suriname, a strain isolated from a 52-year-old man with classical symptoms of ZIKV infection was sequenced and confirmed to belong to the Asian lineage with similarities of nucleotide and amino acid compositions at 99.7% and 99.9%, respectively, compared to strain H/PF/2013 from French Polynesia (35). In Brazil, sequences of isolates from Bahia also indicated an Asian lineage origin with high similarities to those isolated in Sao Paulo e Natal (Brazil) and Suriname (36). ZIKV sequence of an isolate from Mexico showed high similarity to the one isolated in Martinique in 2015, which belongs to the same clade as those isolated in Colombia, Panama, and Brazil (37).

Sequence analyses of isolates from Ecuador (EcEs062_16 and EcEs089_16) revealed that they were more closely related to one strain from Brazil (Paraiba state) than those isolated in Colombia and Peru. It is important to note that Ecuador and Chile are the only countries in South America that do not share a land border with Brazil. The infected patients did not report traveling to Brazil (38). The first case of ZIKV infection in China was imported from Venezuela. The isolate (VE_Ganxian) was sequenced and identified to be Asian lineage closely related to isolates from French Polynesia (Haiti/1225/2014) and Brazil (BeH819966). Some modifications were identified in 11 amino acids in the envelope and 32 amino acids within the complete sequences compared to the Brazilian strain (39).

A complete genome sequence was obtained from a virus isolated in Recife, Brazil, *ZIKV/H. sapiens/Brazil/PE243/2015.* The 5′UTR and 3′UTR of this isolate was also analyzed. A few mutations were identified in NS1, NS3, and NS4, and some were a result of passage in cell culture. Alignment analysis indicated that this strain belonged to the Asian lineage and was closely related to H/PF/2013 from French Polynesia, which matched a Brazilian strain isolated in Sao Paulo (ZikaSPH2015) with similarities of nucleotide and amino acid compositions at 99.9% and 99.97%, respectively. There was no association between one specific modification and the clinical severity of the case (Figures 7.3 and 7.4). A comparison with all sequences available at that time revealed that this strain had a conserved 5′UTR (Figures 7.5 and 7.6) (40). Another study analyzing the 3′UTR identified a subgenomic flavivirus RNA in cells infected with a ZIKV isolate that antagonized IFN-I response. This study identified numerous modifications in the 3′UTR (U42C, C257U, U258C and A266G, U424G, U426C, and C427U) that might be involved with the neurological disorders observed in recent ZIKV outbreaks (41). Unique amino acids have been identified in ZIKV strains isolated from America, which are not present in Asian strains isolated in South Pacific (42). Different modifications have also been identified between Asian/American and African lineages (43). Additionally, translation of the viral proteins is biased with a preference for purine but no difference has been identified among the strains isolated in different countries (44, 45).

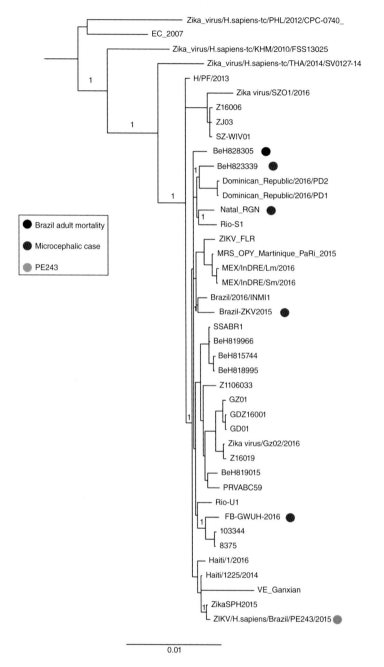

Figure 7.3 Bayesian maximum clade credibility tree generated from coding sequence data. Bayesian posterior probabilities are given at nodes of importance. Isolates implicated in diseases are highlighted. EC_2007 refers to the epidemic consensus sequence generated from the Yap Island outbreak in 2007 (EU545988). *Source:* Figure and legend extracted from Donald et al. (2016) (40).

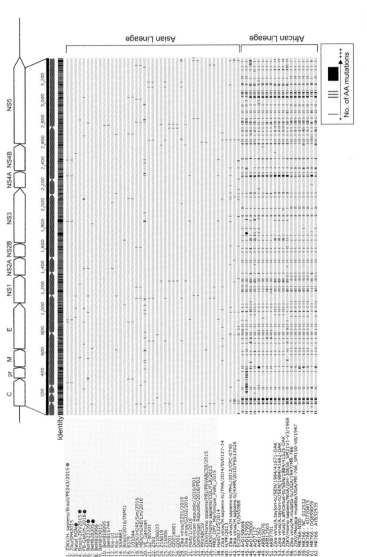

Figure 7.4 Comparison of protein coding region for African and Asian ZIKV lineage. The mean pairwise identity of all pairs at a given position is indicated by the identity bar; light gray denotes 100% pairwise identity, dark gray highlights positions possessing less than 100% pairwise identity. Positions and quantities of amino acid substitutions are indicated by black bands within gray sequence bars. Sequences 1–37, highlighted gray, correspond to the outbreak originating in 2015 in South America. *Source:* Figure and legend extracted from Donald et al. (2016) (40).

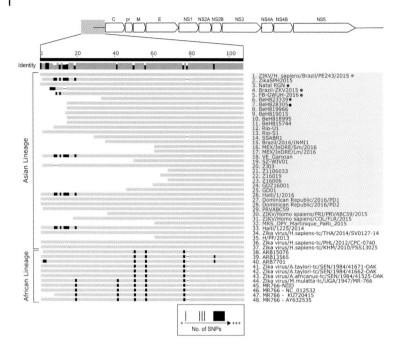

Figure 7.5 Comparison of the 5′UTR nucleotide sequences of Asian and African ZIKV isolates. The mean pairwise identity of all pairs at a given position is indicated by the identity bar; lilac is indicative of 100% pairwise identity, dark gray highlights positions possessing <100% pairwise identity. Positions and quantities of single nucleotide polymorphisms (SNPs) are represented as black bands within gray sequence bars. Sequences 1–32, highlighted gray, correspond to the outbreak originating in 2015 in Brazil. *Source:* Figure and legend extracted from Donald et al. (2016) (40).

To associate sequence variations of different isolates with cell biology, pathology, and the development of microcephaly—which has only been found in a few regions so far—remains a challenge. Studies with animal models have used different strains but there is no dramatic difference in the phenotype so far, particularly for those conducted in immuno-compromised mice (46–53).

For nonhuman primate models, the only association between congenital brain malformation and ZIKV infection was obtained when ZIKV strains recently isolated from Brazil were used (54). There were huge variations among the doses of ZIKV used in animal studies. The injection sites were often different. Thus, differences in all methodologies must be considered before making any conclusion on the pathogenicity of the

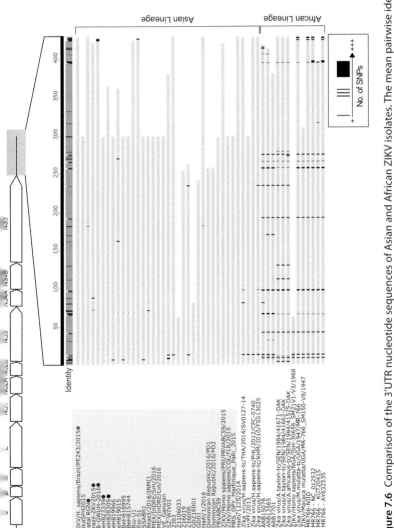

Figure 7.6 Comparison of the 3'UTR nucleotide sequences of Asian and African ZIKV isolates. The mean pairwise identity of all pairs at a given position is indicated by the identity bar: light gray is indicative of 100% pairwise identity, dark gray highlights positions possessing less than 100% pairwise identity. Sequences 1–32, highlighted gray, correspond to the outbreak originating in 2015 in Brazil. *Source:* Figure and legend extracted from Donald et al. (2016) (40).

lineage or strain. At this point, no nonhuman primate study has been simultaneously performed with different lineages. Hence, it is unclear if the Brazilian strain is the only one that can cause microcephaly.

Interestingly, alignment analysis of the strains recently isolated in Brazil, which were probably introduced in one single event, revealed strong sequence variability in a short period of circulation, implicating dynamic modifications of the virus. It is unclear if multiple introductions of ZIKV would result in more genetic diversity. Since strains from other American countries show frequent sequence variations, it is likely that none of the ZIKV strains has achieved fitness yet (55).

References

1 Dick GW. 1952. Zika virus. II. Pathogenicity and physical properties. Trans R Soc Trop Med Hyg **46**:521–534.
2 Dick GW, Kitchen SF, Haddow AJ. 1952. Zika virus. I. Isolations and serological specificity. Trans R Soc Trop Med Hyg **46**:509–520.
3 Weinbren MP, Williams MC. 1958. Zika virus: further isolations in the Zika area, and some studies on the strains isolated. Trans R Soc Trop Med Hyg **52**:263–268.
4 Haddow AJ, Williams MC, Woodall JP, Simpson DI, Goma LK. 1964. Twelve Isolations of Zika Virus from Aedes (Stegomyia) Africanus (Theobald) Taken in and above a Uganda Forest. Bull World Health Organ **31**:57–69.
5 Haddow AD, Schuh AJ, Yasuda CY, Kasper MR, Heang V, Huy R, Guzman H, Tesh RB, Weaver SC. 2012. Genetic characterization of Zika virus strains: geographic expansion of the Asian lineage. PLoS Negl Trop Dis **6**:e1477.
6 Heang V, Yasuda CY, Sovann L, Haddow AD, Travassos da Rosa AP, Tesh RB, Kasper MR. 2012. Zika virus infection, Cambodia, 2010. Emerg Infect Dis **18**:349–351.
7 Baronti C, Piorkowski G, Charrel RN, Boubis L, Leparc-Goffart I, de Lamballerie X. 2014. Complete coding sequence of zika virus from a French polynesia outbreak in 2013. Genome Announc **2**: e00500-00514.
8 Campos GS, Bandeira AC, Sardi SI. 2015. Zika Virus Outbreak, Bahia, Brazil. Emerg Infect Dis **21**:1885–1886.
9 Cao-Lormeau VM, Roche C, Teissier A, Robin E, Berry AL, Mallet HP, Sall AA, Musso D. 2014. Zika virus, French polynesia, South pacific, 2013. Emerg Infect Dis **20**:1085–1086.

10 Oehler E, Watrin L, Larre P, Leparc-Goffart I, Lastere S, Valour F, Baudouin L, Mallet H, Musso D, Ghawche F. 2014. Zika virus infection complicated by Guillain-Barre syndrome--case report, French Polynesia, December 2013. Euro Surveill **19**.

11 Mlakar J, Korva M, Tul N, Popovic M, Poljsak-Prijatelj M, Mraz J, Kolenc M, Resman Rus K, Vesnaver Vipotnik T, Fabjan Vodusek V, Vizjak A, Pizem J, Petrovec M, Avsic Zupanc T. 2016. Zika Virus Associated with Microcephaly. N Engl J Med **374**:951–958.

12 Simpson DI. 1964. Zika Virus Infection in Man. Trans R Soc Trop Med Hyg **58**:335–338.

13 Cao-Lormeau V-M, Blake A, Mons S, Lastère S, Roche C, Vanhomwegen J, Dub T, Baudouin L, Teissier A, Larre P, Vial A-L, Decam C, Choumet V, Halstead SK, Willison HJ, Musset L, Manuguerra J-C, Despres P, Fournier E, Mallet H-P, Musso D, Fontanet A, Neil J, Ghawché F. Guillain-Barre Syndrome outbreak associated with Zika virus infection in French Polynesia: a case-control study. The Lancet **387**:1531–1539.

14 Macnamara FN. 1954. Zika virus: a report on three cases of human infection during an epidemic of jaundice in Nigeria. Trans R Soc Trop Med Hyg **48**:139–145.

15 Bearcroft WG. 1956. Zika virus infection experimentally induced in a human volunteer. Trans R Soc Trop Med Hyg **50**:442–448.

16 Wikan N, Smith DR. 2017. First published report of Zika virus infection in people: Simpson, not MacNamara. Lancet Infect Dis **17**:15–17.

17 Lanciotti RS, Kosoy OL, Laven JJ, Velez JO, Lambert AJ, Johnson AJ, Stanfield SM, Duffy MR. 2008. Genetic and serologic properties of Zika virus associated with an epidemic, Yap State, Micronesia, 2007. Emerg Infect Dis **14**:1232–1239.

18 Yun SI, Song BH, Frank JC, Julander JG, Polejaeva IA, Davies CJ, White KL, Lee YM. 2016. Complete Genome Sequences of Three Historically Important, Spatiotemporally Distinct, and Genetically Divergent Strains of Zika Virus: MR-766, P6-740, and PRVABC-59. Genome Announc **4**.

19 Kuno G, Chang GJ. 2007. Full-length sequencing and genomic characterization of Bagaza, Kedougou, and Zika viruses. Arch Virol **152**:687–696.

20 Chambers TJ, Halevy M, Nestorowicz A, Rice CM, Lustig S. 1998. West Nile virus envelope proteins: nucleotide sequence analysis of strains differing in mouse neuroinvasiveness. J Gen Virol **79** (Pt 10):2375–2380.

21 Beasley DW, Whiteman MC, Zhang S, Huang CY, Schneider BS, Smith DR, Gromowski GD, Higgs S, Kinney RM, Barrett AD. 2005. Envelope protein glycosylation status influences mouse neuroinvasion phenotype of genetic lineage 1 West Nile virus strains. J Virol **79**:8339–8347.

22 Taylor RM. 1952. Studies on certain viruses isolated in the tropics of Africa and South America; their growth and behavior in the embryonated hen egg. J Immunol **68**:473–494.

23 Ladner JT, Wiley MR, Prieto K, Yasuda CY, Nagle E, Kasper MR, Reyes D, Vasilakis N, Heang V, Weaver SC, Haddow A, Tesh RB, Sovann L, Palacios G. 2016. Complete Genome Sequences of Five Zika Virus Isolates. Genome Announc **4**.

24 Ellison DW, Ladner JT, Buathong R, Alera MT, Wiley MR, Hermann L, Rutvisuttinunt W, Klungthong C, Chinnawirotpisan P, Manasatienkij W, Melendrez MC, Maljkovic Berry I, Thaisomboonsuk B, Ong-Ajchaowlerd P, Kaneechit W, Velasco JM, Tac-An IA, Villa D, Lago CB, Roque VG, Jr., Plipat T, Nisalak A, Srikiatkhachorn A, Fernandez S, Yoon IK, Haddow AD, Palacios GF, Jarman RG, Macareo LR. 2016. Complete Genome Sequences of Zika Virus Strains Isolated from the Blood of Patients in Thailand in 2014 and the Philippines in 2012. Genome Announc **4**.

25 Zhu Z, Chan JF, Tee KM, Choi GK, Lau SK, Woo PC, Tse H, Yuen KY. 2016. Comparative genomic analysis of pre-epidemic and epidemic Zika virus strains for virological factors potentially associated with the rapidly expanding epidemic. Emerg Microbes Infect **5**:e22.

26 Simonin Y, Loustalot F, Desmetz C, Foulongne V, Constant O, Fournier-Wirth C, Leon F, Moles JP, Goubaud A, Lemaitre JM, Maquart M, Leparc-Goffart I, Briant L, Nagot N, Van de Perre P, Salinas S. 2016. Zika Virus Strains Potentially Display Different Infectious Profiles in Human Neural Cells. EBioMedicine **12**:161–169.

27 Hamel R, Ferraris P, Wichit S, Diop F, Talignani L, Pompon J, Garcia D, Liegeois F, Sall AA, Yssel H, Misse D. 2017. African and Asian Zika virus strains differentially induce early antiviral responses in primary human astrocytes. Infect Genet Evol **49**:134–137.

28 Cardoso CW, Paploski IA, Kikuti M, Rodrigues MS, Silva MM, Campos GS, Sardi SI, Kitron U, Reis MG, Ribeiro GS. 2015. Outbreak of Exanthematous Illness Associated with Zika, Chikungunya, and Dengue Viruses, Salvador, Brazil. Emerg Infect Dis **21**:2274–2276.

29 Musso D. 2015. Zika Virus Transmission from French Polynesia to Brazil. Emerg Infect Dis **21**:1887.

30 Zanluca C, Melo VC, Mosimann AL, Santos GI, Santos CN, Luz K. 2015. First report of autochthonous transmission of Zika virus in Brazil. Mem Inst Oswaldo Cruz **110**:569–572.

31 Araujo LM, Ferreira ML, Nascimento OJ. 2016. Guillain-Barre syndrome associated with the Zika virus outbreak in Brazil. Arq Neuropsiquiatr **74**:253–255.

32 Brasil P, Calvet GA, Siqueira AM, Wakimoto M, de Sequeira PC, Nobre A, Quintana Mde S, Mendonca MC, Lupi O, de Souza RV, Romero C, Zogbi H, Bressan Cda S, Alves SS, Lourenco-de-Oliveira R, Nogueira RM, Carvalho MS, de Filippis AM, Jaenisch T. 2016. Zika Virus Outbreak in Rio de Janeiro, Brazil: Clinical Characterization, Epidemiological and Virological Aspects. PLoS Negl Trop Dis **10**:e0004636.

33 Liu H, Shen L, Zhang XL, Li XL, Liang GD, Ji HF. 2016. From discovery to outbreak: the genetic evolution of the emerging Zika virus. Emerg Microbes Infect **5**:e111.

34 Piorkowski G, Richard P, Baronti C, Gallian P, Charrel R, Leparc-Goffart I, de Lamballerie X. 2016. Complete coding sequence of Zika virus from Martinique outbreak in 2015. New Microbes New Infect **11**:52–53.

35 Enfissi A, Codrington J, Roosblad J, Kazanji M, Rousset D. 2016. Zika virus genome from the Americas. Lancet **387**:227–228.

36 Giovanetti M, Faria NR, Nunes MR, de Vasconcelos JM, Lourenco J, Rodrigues SG, Vianez JL, Jr., da Silva SP, Lemos PS, Tavares FN, Martin DP, do Rosario MS, Siqueira IC, Ciccozzi M, Pybus OG, de Oliveira T, Alcantara LCJ. 2016. Zika virus complete genome from Salvador, Bahia, Brazil. Infect Genet Evol **41**:142–145.

37 Diaz-Quinonez JA, Pena-Alonso R, Mendieta-Condado E, Garces-Ayala F, Gonzalez-Duran E, Escobar-Escamilla N, Vazquez-Pichardo M, Torres-Rodriguez Mde L, Nunez-Leon A, Torres-Longoria B, Lopez-Martinez I, Ruiz-Matus C, Kuri-Morales P, Ramirez-Gonzalez JE. 2016. Complete Genome Sequence of Zika Virus Isolated in Mexico, 2016. Genome Announc **4**.

38 Marquez S, Carrera J, Pullan ST, Lewandowski K, Paz V, Loman N, Quick J, Bonsall D, Powell R, Theze J, Pybus OG, Klenerman P, Eisenberg J, Coloma J, Carroll MW, Trueba G, Logue CH. 2017. First Complete Genome Sequences of Zika Virus Isolated from Febrile Patient Sera in Ecuador. Genome Announc **5**.

39 Liu L, Wu W, Zhao X, Xiong Y, Zhang S, Liu X, Qu J, Li J, Nei K, Liang M, Shu Y, Hu G, Ma X, Li D. 2016. Complete Genome Sequence of Zika Virus from the First Imported Case in Mainland China. Genome Announc **4**.

40 Donald CL, Brennan B, Cumberworth SL, Rezelj VV, Clark JJ, Cordeiro MT, Freitas de Oliveira Franca R, Pena LJ, Wilkie GS, Da Silva Filipe A, Davis C, Hughes J, Varjak M, Selinger M, Zuvanov L, Owsianka AM, Patel AH, McLauchlan J, Lindenbach BD, Fall G, Sall AA, Biek R, Rehwinkel J, Schnettler E, Kohl A. 2016. Full Genome Sequence and sfRNA Interferon Antagonist Activity of Zika Virus from Recife, Brazil. PLoS Negl Trop Dis **10**:e0005048.

41 Yokoyama S, Starmer WT. 2017. Possible Roles of New Mutations Shared by Asian and American Zika Viruses. Mol Biol Evol **34**:525–534.

42 Kochakarn T, Kotanan N, Kumpornsin K, Loesbanluechai D, Thammasatta M, Auewarakul P, Wilairat P, Chookajorn T. 2016. Comparative genome analysis between Southeast Asian and South American Zika viruses. Asian Pac J Trop Med **9**:1048–1054.

43 Maurer-Stroh S, Mak TM, Ng YK, Phuah SP, Huber RG, Marzinek JK, Holdbrook DA, Lee RT, Cui L, Lin RT. 2016. South-east Asian Zika virus strain linked to cluster of cases in Singapore, August 2016. Euro Surveill **21**.

44 van Hemert F, Berkhout B. 2016. Nucleotide composition of the Zika virus RNA genome and its codon usage. Virol J **13**:95.

45 Butt AM, Nasrullah I, Qamar R, Tong Y. 2016. Evolution of codon usage in Zika virus genomes is host and vector specific. Emerg Microbes Infect **5**:e107.

46 Huang WC, Abraham R, Shim BS, Choe H, Page DT. 2016. Zika virus infection during the period of maximal brain growth causes microcephaly and corticospinal neuron apoptosis in wild type mice. Sci Rep **6**:34793.

47 Manangeeswaran M, Ireland DD, Verthelyi D. 2016. Zika (PRVABC59) Infection Is Associated with T cell Infiltration and Neurodegeneration in CNS of Immunocompetent Neonatal C57Bl/6 Mice. PLoS Pathog **12**:e1006004.

48 Miner JJ, Cao B, Govero J, Smith AM, Fernandez E, Cabrera OH, Garber C, Noll M, Klein RS, Noguchi KK, Mysorekar IU, Diamond MS. 2016. Zika Virus Infection during Pregnancy in Mice Causes Placental Damage and Fetal Demise. Cell **165**:1081–1091.

49 Zhang NN, Tian M, Deng YQ, Hao JN, Wang HJ, Huang XY, Li XF, Wang YG, Zhao LZ, Zhang FC, Qin CF. 2016. Characterization of the contemporary Zika virus in immunocompetent mice. Hum Vaccin Immunother **12**:3107–3109.

50 Pardy RD, Rajah MM, Condotta SA, Taylor NG, Sagan SM, Richer MJ. 2017. Analysis of the T Cell Response to Zika Virus and Identification of a Novel CD8+ T Cell Epitope in Immunocompetent Mice. PLoS Pathog **13**:e1006184.

51 Vermillion MS, Lei J, Shabi Y, Baxter VK, Crilly NP, McLane M, Griffin DE, Pekosz A, Klein SL, Burd I. 2017. Intrauterine Zika virus infection of pregnant immunocompetent mice models transplacental transmission and adverse perinatal outcomes. Nat Commun **8**:14575.

52 Wang S, Hong S, Deng YQ, Ye Q, Zhao LZ, Zhang FC, Qin CF, Xu Z. 2017. Transfer of convalescent serum to pregnant mice prevents Zika virus infection and microcephaly in offspring. Cell Res **27**:158–160.

53 Cugola FR, Fernandes IR, Russo FB, Freitas BC, Dias JL, Guimaraes KP, Benazzato C, Almeida N, Pignatari GC, Romero S, Polonio CM, Cunha I, Freitas CL, Brandao WN, Rossato C, Andrade DG, Faria Dde P, Garcez AT, Buchpigel CA, Braconi CT, Mendes E, Sall AA, Zanotto PM, Peron JP, Muotri AR, Beltrao-Braga PC. 2016. The Brazilian Zika virus strain causes birth defects in experimental models. Nature **534**:267–271.

54 Adams Waldorf KM, Stencel-Baerenwald JE, Kapur RP, Studholme C, Boldenow E, Vornhagen J, Baldessari A, Dighe MK, Thiel J, Merillat S, Armistead B, Tisoncik-Go J, Green RR, Davis MA, Dewey EC, Fairgrieve MR, Gatenby JC, Richards T, Garden GA, Diamond MS, Juul SE, Grant RF, Kuller L, Shaw DW, Ogle J, Gough GM, Lee W, English C, Hevner RF, Dobyns WB, Gale M, Jr., Rajagopal L. 2016. Fetal brain lesions after subcutaneous inoculation of Zika virus in a pregnant nonhuman primate. Nat Med **22**:1256–1259.

55 Shi W, Zhang Z, Ling C, Carr MJ, Tong Y, Gao GF. 2016. Increasing genetic diversity of Zika virus in the Latin American outbreak. Emerg Microbes Infect **5**:e68.

8

ZIKV-Host Interactions

Virus infection usually causes global changes in the host cells. On one hand, viral infections induce host innate immune responses, which is a mechanism for the host cells to defend the invading viruses. On the other hand, viruses have often evolved specific mechanisms to counter host defense mechanisms, which are essential for efficient viral infection and replication. Following the recent ZIKV outbreak, several studies have systematically profiled transcriptomes and proteomes of the ZIKV-infected cells (1–5). These works have shown that several host functions or signaling pathways are drastically altered as a result of ZIKV infection. These functions include, but are not limited to, innate immune response, apoptosis and cell death, autophagy, cell cycle and mitosis, DNA replication and DNA repair, and protein synthesis and transport, among others. This chapter will review the most recent findings in the cellular functions and pathways altered during ZIKV infection.

8.1 Systematic Studies to Identify ZIKV Affected Functions and Pathways

Several recent studies of high-throughput screenings of transcriptome and proteome have provided comprehensive understanding of the biological processes and functions altered during ZIKV infection. Because of the association of microcephaly with ZIKV infection, Tang et al. (2016) performed a global gene expression profiling using a model of ZIKV infection of human inducible pluripotent stem cells (hiPSCs) (2).

Zika Virus and Diseases: From Molecular Biology to Epidemiology, First Edition.
Suzane Ramos da Silva, Fan Cheng and Shou-Jiang Gao.
© 2018 John Wiley & Sons, Inc. Published 2018 by John Wiley & Sons, Inc.

Upon proper induction, hiPSCs could be differentiated into forebrain-specific human neural progenitor cells (hNPCs), which could be further differentiated into cortical neurons (6). Tang et al. (2016) showed that the hPSCs could be infected by ZIKV (MR766 strain, African) with a high rate. Human embryonic stem cells (hESCs) and hiPSCs can also be infected, but the infection rates were low, with reduced expression of pluripotent marker NANOG (2). These findings suggest that hNPCs, a constitutive population of the developing embryonic brain, could be the direct targets of ZIKV infection.

Global transcriptome analyses with this model identified a large number of differentially expressed genes upon viral infection (2). Gene Ontology (GO) analyses revealed a particular enrichment of down-regulated genes in DNA metabolism, particularly those in cell-cycle-related pathways (1, 2). Up-regulated genes were primarily enriched in nucleic acid metabolism regulation, including transcription and catabolic process, as well as protein transport. Cellular compartment subontology further revealed the locations where ZIKV-induced alterations of gene expression occurred (1). Down-regulated genes were highly enriched in centromeric and nuclear sites, the extracellular matrix, exosomes, and mitochondria while up-regulated genes were enriched in the Golgi complex, the cytoplasm, and MOZ (monocytic leukemic zinc-finger protein)/MORF (MOZ-related factor) histone acetyltransferase complex, which play key roles in regulating different developmental programs, including skeletogenesis, neurogenesis and hematopoiesis (1, 7).

By plotting a GO network map using human genome as the background, and GO biological process and immune system terms queried for enrichment, Rolfe et al. (2016) identified eight main networks (1). Four of the networks were associated with immune system response, including immune response, cytokine production, leukocyte activation, and defense response to other organisms. Other enriched networks are nucleic acid metabolic processes, cell cycle, negative regulation of macromolecule biosynthetic process, and cellular macromolecule metabolic process (1).

In another systematic study, Zhang et al. (2016) profiled transcriptomes of hPSCs exposed to Asian ZIKV^M (MR766 strain), African ZIKV^C (FSS13025 strain), and dengue virus (DENV), as well as analyzed virus- and strain-specific molecular signatures associated with ZIKV infection (3). Comparison between DENV and ZIKV^M revealed enrichments of 78, 32, and 26 genes in cell cycle, DNA replication, and DNA repair-related pathways that were down-regulated in both DENV- and ZIKV^{M}-infected cells, respectively, as well as a noticeable enrichment for genes in protein synthesis and apoptosis-related pathways that were up-regulated in both groups. For virus-specific transcriptional regulation, 19 and 8 genes related to DNA replication and replication fork GO terms, respectively, were down-regulated in ZIKV^{M}-infected hPSCs but not in DENV-infected cells. In this study, Zhang et al. (2016) also compared the gene expression profiles of ZIKV^{M}- and ZIKV^{C}-infected hPSCs (3). Gene expression analyses revealed that there was significant overlap of genes with altered expression in infected hPSCs between African and Asian strains. Genes consistently down-regulated upon infection by both ZIKV strains were mainly enriched in DNA replication, cell cycle, and DNA repair while up-regulated genes were enriched in responses to unfolded protein and cell death. Interestingly, the Asian strain induced specific dysregulation of DNA replication and repair genes (e.g., TP53), and the up-regulation of viral response genes in pathways of Toll-like receptor (TLR) signaling and TNF-α signaling in hPSCs (3).

In another study, Garcez et al. (2017) performed proteome and transcriptome analyses with cells of mock- and ZIKV-infected neurospheres (4). Proteomics analysis revealed that human neurosphere infected by ZIKV (Brazilian strain) had down-regulation of proteins involved in organelle localization and regulation, and protein folding, as well as down-regulation of translation and cell cycle-related processes. By combining proteomics and mRNA transcriptional profiling, it was shown that over 500 proteins and genes were differentially expressed. An interactome map was constructed, which predicted the functional networks related to viral replication, cell cycle arrest, chromosomal instability as well as decline of the neurogenic program (4).

The human placenta is a chimeric organ containing both maternal and fetal structure: the maternal decidua, the specialized endometrium of pregnancy that constitutes the uterine implantation site, and the fetus-derived chorionic villi (8). The placenta is well known to be armed with both physical and innate immune barriers against invading pathogens (8, 9). Weisblum et al. (2017) demonstrated that ZIKV (PRVABC59 strain) can effectively replicate in both maternal and fetal facets of the placenta. In a genome-wide transcriptome analysis of ZIKV-infected versus mock-infected tissues (5), it was shown that ZIKV infection significantly upregulated some innate immunity genes In the decidual tissue, particularly genes related to antiviral interferon signaling pathways (mostly type I and type III IFNs) though ZIKV did not activate the decidual tissue immune cell response. In contrast, responses to chorionic villus ZIKV infection were distinctively enriched for apoptosis, cell death, and necrosis molecular functions (5).

The systematic, genome-wide gene expression analyses have so far revealed the molecular signatures associated with ZIKV infection. In the following sections of this chapter, each individual biological processes or functions that are affected by ZIKV infection will be described in details.

8.2 Induction and Dysregulation of Innate Immune Responses during ZIKV Infection

8.2.1 Host Innate Immunity against Infections of Flaviviruses

To combat virus infection, the host encodes pattern recognition receptors (PRRs) residing within the cytoplasm and endosomal vesicles to rapidly control viral replication and prevent virus spread by recognizing non-self nucleic acids termed pathogen-associated molecular patterns (PAMPs), hence triggering antiviral responses (10). However, many pathogenic viruses use distinct strategies to evade the innate immune responses (11).

Two classes of PRRs, the retinoid-inducible gene I (RIG-I)-like receptors (RLRs) and the TLRs, are essential for responding to flavivirus infection (12). Both RLR and TLR receptors are implicated in the induction of an immune response against flaviviruses. The activation of these PRRs leads to the production of type I IFNs and other pro-inflammatory cytokines [reviewed in (10, 12)]. Flaviviruses have also evolved a variety of strategies to modulate and/or evade the immune responses of host cells [reviewed in (12–14)].

RLRs and melanoma-differentiation-associated gene 5 (MDA5) are cytosolic innate immune PRRs that sense double-stranded RNA (dsRNA), a replication intermediate of RNA viruses (Figure 8.1) (15). RIG-I engages short dsRNA containing a 5′-triphosphate and uridine- or adenosine-rich motifs while MDA5 does not require 5′-triphosphate but preferentially engages long dsRNA (15, 16). Binding of viral PAMPs to the helicase domains of RLRs induces conformational changes, which allow these RLRs to interact with the adaptor protein

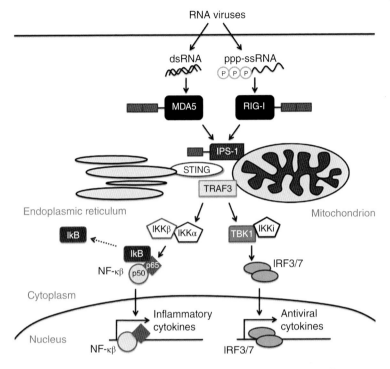

Figure 8.1 RIG-I-like receptors (RLRs) serve as cytosolic sensors for viral RNA to initiate anti-viral and inflammatory cell responses. MDA5 recognizes dsRNA while RIG-I perceives short ssRNAs or dsRNAs with 5′ triphosphate ends (ppp-ssRNA). Interaction with viral RNAs recruits downstream effector molecules (e.g., IPS-1, TRAF 3, and/or STING), thus activating IKK-related kinases, TBK1, and IKKi. These kinases further induce transcription factors NF-κB and IRF3/7 to activate transcription of anti-viral and inflammatory cytokines. *Source:* Extracted from Furr et al. (2012) (131).

IPS-1/MAVS/Cardif. These interactions initiate a signaling cascade, resulting in the activation of transcription factors such as IFN regulatory factors (IRF) 3 and 7, and NF-κB, which are required for the production of type I IFN, proinflammatory cytokines, and expression of IFN-stimulated genes (ISGs), as well as the establishment of an antiviral state within the cells.

TLR family members are evolutionarily conserved transmembrane molecules that are expressed on the cell surface or within endocytic vesicles in a cell type-dependent manner (17, 18). TLR3 and TLR7, which reside within endosomal vesicles, function as broad sensors of dsRNA and single-stranded RNA (ssRNA), respectively. However, TLR7 response to ssRNA is enhanced by higher order structures within viral RNA (19–21). Upon binding to PAMP RNA, the adaptor protein MyD88 mediates the signaling pathways of TLR7 while TLR3 utilizes TIR-domain-containing adaptor-inducing IFN-β (TRIF) (17, 18). Similar to the RLRs, stimulation of the TLR pathways results in the production of IFN-α/β and inflammatory cytokines, which, in turn, stimulates maturation of dendritic cells (DCs) and the establishment of an antiviral response (22, 23).

8.2.2 Activation of Host Innate Immune Responses by ZIKV in Various Cell Models

Similar to other flaviviruses, ZIKV infection induces innate antiviral response in the host. Using primary human skin fibroblasts as a model, Hamel et al. (2015) first showed that the expression levels of RIG-I, MDA5, and TLR3 were upregulated at as early as 6 hours post-infection (h.p.i.), with maximal mRNA levels detected at 48 h.p.i., in ZIKV-infected cells (24). Using a human qPCR array covering 84 human antiviral genes, it was shown that ZIKV infection enhanced the expression of not only IFN-α and IFN-β but also several ISGs, including OAS2, ISG15, and MX1. IRF3 and IRF7 are generally considered as the master regulators for the induction of type I IFN and ISGs (25, 26). Consistently, the mRNA level of IRF7 was increased in ZIKV-infected cells. In addition, the expression of CXCR3 ligand CXCL10, the inflammatory antiviral chemokine CCL5, as well as inflammasome components AIM2 and interleukin-1β (IL-1β) was also induced following ZIKV infection. Besides, inhibition of TLR3 expression, but not other PRRs, led to a potent increase in the viral RNA copy number at 48 h.p.i., underscoring the importance of TLR3 in the induction of the anti-ZIKV response. Lastly, pre-treatment of primary skin fibroblasts with IFN-α, IFN-β, or IFN-γ prior to ZIKV infection at multiplicity of infection (MOI) of 1 strongly and dose-dependently inhibited progeny virus production with similar efficacies (24).

In another study, Frumence et al. (2016) used lung epithelial A549 cells to evaluate their innate immune response to ZIKV (PF-25013-18 strain) infection (27). It has been shown that the IFN-induced proteins with tetratricopeptide repeats (IFIT) possess antiviral activities against flaviviruses, and that DENV infection induced IFIT expression in A549 cells (28–30). At 18 h.p.i., ZIKV infection stimulated the transcription of both IFIT1 (ISG56) and IFIT2 (ISG54) genes and their transcripts increased over the course of the infection. As expected, the amount of IRF3 and IRF7 transcripts were increased approximately 15-fold at 24 and 48 h.p.i. Furthermore, it was observed that the induction of IL-8 occurred within 24 hours of infection while those of IL-6, IL-1β, and the monocyte chemoattractant protein-1 (MCP1) occurred at 48 h.p.i. ZIKV-infected A549 cells secreted IFN-β and IL-8, and to a lesser extent, IL-6 (27). Similar to what was observed in primary human skin fibroblasts (24), IFN-β pre-treatment resulted in a drastic reduction of viral replication examined at 48 h.p.i. in A549 cells without any obvious effect on cell viability (27).

Using human cervical carcinoma cells HeLa, Savidis et al. (2016) demonstrated that IFN-inducible transmembrane proteins (IFITMs), particularly IFITM3 and IFITM1, inhibited ZIKV virus replication (31). Extensive studies have shown that IFITMs can inhibit the replication of a wide range of pathogenic viruses including all flaviviruses tested to date. These include WNV, DENV, and reporter viruses carrying the envelope of the Omsk hemorrhagic fever virus (32–40). Loss- and gain-of-function experiments suggested that IFITM3 and IFITM1 inhibited the replication of both prototype (MR766) and contemporary (FSS13025) ZIKV strains with IFITM3 possessing a higher inhibitory potency. IFITM3 inhibited early stages of ZIKV replication though the exact mechanism remained unknown (31). Given that the results of orthologous experimental approaches suggest that IFITMs restrict viral replication by blocking fusion-pore formation, and the entry of viral genome and its associated proteins into cytosol (32, 33, 39, 41–43), the inhibition by IFITM3 is speculated to target early stages of ZIKV infection after virus binding but before the cytosolic entry or transcription of viral RNA.

DCs are critical immune sentinel cells, bridging pathogen detection to activation of innate and adaptive antiviral immunity. The interplay between multiple mosquito-borne flaviviruses and DCs has been extensively investigated (44–47). Recent study by Bowen et al. (2017) showed that human DCs supported productive ZIKV infection. African lineage viruses (MR766 and DakAr 41524 strains) displayed more rapid replication kinetics and infection magnitudes than Asian lineage viruses did (P6-740 and PRVABC59 strains) with the African strains also inducing cell death (48). A critical function of DCs is to program virus-specific T cell responses in order to clear the virus-infected cells. By measuring the cell surface expression of co-stimulatory markers (CD80, CD86, and CD40) and MHC class II molecules on DCs, which indicated their ability to prime virus-specific T cell responses (49), it was observed that infection of monocyte-derived DCs (moDCs) with ZIKV P6-740 or DaKar 41524 strain only induced minimal activation while infection with ZIKV PRVABC59 or MR766 strain led to modest activation. As a consequence, the abilities of infected DCs to prime T cell responses may be compromised (48). In addition, infection of DCs with both African and Asian lineage ZIKV strains led to limited secretion of inflammatory cytokines (e.g. IL-6, MCP1, RANTES, and IP-10). Despite strong induction of type I IFN was observed at the transcriptional level, their protein translation was inhibited during ZIKV infection. When the gene expression levels of host RLRs and other host antiviral effectors were examined, upregulation of RLRs (RIG-I, MDA5, and LGP2) and other antiviral effectors (IFIT family members, OAS1, and viperin) were observed in response to both contemporary and historic ZIKV strains. Lastly, treatment of human DCs with RIG-I agonist potently restricted ZIKV replication while type I IFN treatment only resulted in modest but non-significant decreases in viral replication. All of these findings suggest that ZIKV selectively inhibits type I IFN translation but does not affect the other host antiviral effectors (48).

During mammalian pregnancy, the placenta acts as a physical and immunological barrier between the maternal and fetus compartments, and protects the developing fetus from the vertical transmission of viruses. The existing evidence of the causal link between ZIKV infection and microcephaly suggests that ZIKV may bypass the placenta barrier to reach the fetus (50, 51). Bayer et al. (2016) assessed the ability of ZIKV (MR766 and FSS13025 strains) to replicate in primary human trophoblast (PHT) cells, which are the barrier cells of placenta, a panel of trophoblast-derived cell lines, including BeWo, JEG-3 and JAR choriocarcinoma cells, and the extravillous trophoblast cell line HTR8/SVneo (9, 52). The investigation found that BeWo, JEG-3, JAR, and HTR8/SVneo cells supported infection of both ZIKV strains with BeWo cells less susceptible to ZIKV infection than the other trophoblast-derived cell lines. In contrast, no ZIKV replication was detected in PHT cells, which was consistent with previous observations that PHT cells resist infection by diverse DNA and RNA viruses (9, 53). In addition, Bayer et al. (2016) found that conditioned medium (CM) from uninfected PHT cells protected non-placental cells from ZIKV infection while CM from other trophoblast cell lines did not. The ELISA data indicated that PHT cells constitutively released type III interferon IFNλ1, which functioned in a paracrine and autocrine manner to protect trophoblast and non-trophoblast cells from ZIKV infection (9). Another study done by Quicke et al. (2016) showed that human primary placental macrophages, called Hofbauer cells (HCs), are permissive to productive infection by ZIKV (PRVABC59 strain) (54). Upon infection, HCs were modestly activated and produced IFN-α and other pro-inflammatory cytokines, including RSAD2, IFIT1, IFIT2, IFIT3, and OAS1 (54).

Hanners et al. (2016) also evaluated the immunogenicity of ZIKV (PRVABC59 strain) in primary human fetal neural progenitor cells (fNPCs) (55). Two fNPC lines were acquired at weeks 16–19 of gestation, a time-frame of fetal development correlating with risk for severe fetal neurologic complications after ZIKV exposure (55, 56). The results showed that ZIKV could replicate in fNPCs and induced cell death through apoptosis. To further evaluate the immunogenicity of ZIKV, an antibody-based assay was used to simultaneously detect over 102 human cytokines and chemokines. Surprisingly, no significant difference was observed between mock- and ZIKV-infected fNPCs cultures though multiple cytokines and growth factors were secreted by the cultures. ELISA and RT-qPCR analyses also confirmed the array data showing that fNPCs were refractory to the induction of neuro-inflammatory or neuroprotective cytokines in response to ZIKV infection (55). This study indicates that, following an initial cytotoxic phase, fNPCs support a persistent ZIKV infection.

8.2.3   Suppression of Host Innate Immunity by ZIKV

While several viral proteins of DENV and WNV can inhibit either IFN induction or IFN signaling, the NS5 protein of all human flaviviruses is a potent and specific antagonist of IFN signaling (57–68). Interestingly, flavivirus NS5 proteins exhibit their inhibitory activities against IFN signaling by different mechanisms. DENV NS5 recruits the host E3 ubiquitin ligase UBR4 to degrade STAT2 (68); YFV NS5 binds to STAT2 and prevents STAT2 binding to IFN-stimulated responsive element (ISRE) that are present upstream of ISGs (63); and WNV NS5 prevents IFNAR1 surface expression by targeting the host protein prolidase (66).

The mechanism by which ZIKV antagonized IFN signaling was first revealed by Grant et al. (2016) (69). In this study, the expression of ZIKV NS5 (strain KJ776791.1 of French Polynesia) prevented the activation of ISG54, an IFN-inducible reporter gene, in 293 T cells treated with type I IFN, through direct binding to STAT2, but not to STAT1. Furthermore, Vero cells infected with two different ZIKV strains (KJ776791.1 and MR766) exhibited dramatically reduced STAT2 levels due to proteasome-mediated degradation. The only other flavivirus known to target STAT2 for degradation in infected cells is DENV (61, 68). However, in contrast to DENV, which co-opts the host E3 ubiquitin ligase UBR4 to induce STAT2 degradation, ZIKV infection reduced STAT2 levels independent of the expression of UBR4 (69). Additionally, Grant et al. (2016) also reported that ZIKV antagonism of IFN signaling might be species dependent; ZIKV infection does not affect STAT2 levels in primary mouse embryonic fibroblasts (MEFs). Taken together, ZIKV NS5 inhibits IFN-mediated signaling transduction through a mechanism involving STAT2 binding and most likely STAT2 protein degradation, which shares similarities with that of DENV NS5 (61, 68, 69).

The investigation by Bowen et al. (2017) revealed a similar ZIKV antagonism strategy for type I IFN signaling using A549 cell line and moDCs (48). Infection with ZIKV in the absence of IFN-β treatment induced only minimal STAT1 phosphorylation and low level of STAT1 phosphorylation while treatment of infected cells with IFN-β increased the phosphorylation levels of STAT1, and to a lesser extent, STAT2 but to notably lower levels compared to uninfected cells treated with IFN-β. These data suggest that ZIKV antagonizes the phosphorylation of both STAT1 and STAT2 in human DCs (48), which is, to some extent, different from the findings of Grant et al. (2016) (69).

8.3 Induction of Cell Death and Apoptosis by ZIKV

The aim of a virus is to infect and propagate. In order to maximize this goal, viruses often manipulate the cell survival pathway (70). A large variety of viruses from different families have life cycles that intertwine with critical pathways involved in cell death and survival [review in (71)].

Apoptosis, the most studied form of programmed cell death (PCD), was first reported by Walter Flemming in 1885 (72). In contrast to necrosis, which is a form of cell injury resulting in the premature death of cells in the living tissue (73), apoptosis is a highly regulated process that confers advantages during the life cycle of an organism. Apoptosis can be initiated through two distinct pathways, the extrinsic (death receptor) and intrinsic (mitochondrial) pathways (Figure 8.2) (74). The extrinsic pathway is triggered by the binding of ligands, including tumor necrosis factor-α (TNF-α), FasL, TRAIL, to the death receptors, such as TNF-RI, Fas/CD95, DR3, TRAIL R1/DR4, or TRAIL R2/DR5 on the cell surface (75–77). This binding leads to the trimerization of the membrane receptor, followed by the downstream activation of death-inducing signaling complex (DISC). As a consequence, the protein complex initiates cleavage and activation of caspase-8, which proceeds to trigger pro-caspase-3, the penultimate enzyme for execution of the apoptotic process (78). The intrinsic apoptosis is proximately activated by damage to mitochondria, which releases cytochrome C and apoptosis-activating factor from mitochondria. Cytochrome C recruits Apaf–1 and pro-caspase-9 to compose the apoptosome. By means of this complex, caspase-3 is activated and, as in extrinsically activated apoptosis, caspases 3 and 7 destroy the substructure of the cell (74, 79).

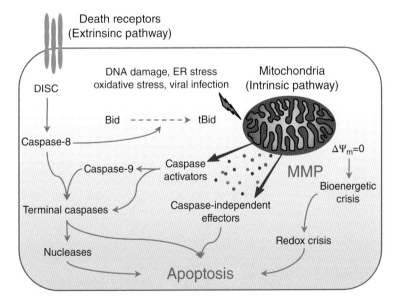

Figure 8.2 The intrinsic (or mitochondrial) and extrinsic (death receptor) pathways of apoptosis. *Source:* Extracted from Galluzzi et al. (2008) (132).

Neurogenesis is the key process by which neurons are generated from neural stem cells (NSCs) and neural progenitor cells (NPCs). Several independent studies have shown that infection by both African and Asian ZIKV leads to cell death in NSCs and NPCs (2–4, 55, 80–84). At 24 h.p.i., apoptotic cells with condensed fragmented nuclei and numerous autophagic vacuoles in the cytoplasm were detected (80). Caspase-8 (extrinsic pathways), caspase-9 (intrinsic pathway), and caspases 3/7 were significantly activated at 24 h.p.i., with the activation of caspases 3/7 activities even more intense after 48 h of infection. Treatment of hNPCs with pan-caspase inhibitor Z-VAD-FMK rescued the reduction of hNPC population induced by ZIKV while in the meantime, maintained a progressive increase of viral genomic replication (80).

In addition, Ghouzzi et al. (2016) found that phosphorylation of p53 Ser15 residue, an established marker of p53 functional activation, is significantly increased by ZIKV infection (81). As expected, the expression of p53 downstream targets, such as AEN, CDKN1A, and CCNG1, was also significantly increased in the infected cells. Interestingly, in this study, the correlation between apoptosis and viral load in infected cells was not evident, as ~60.0% of caspase-3-positive cells were negative for ZIKV viral protein expression (81). Flaviviruses are known to induce genotoxic stress in infected cells, which is one of the main stimuli for p53 activation (85, 86). By quantifying nuclear foci of phosphorylated H2Ax histone (γH2Ax), a well-established marker of DNA double-strand breaks (87), it was showed that ZIKV infection led to increased DNA damage regardless of whether the cells had detectable viral antigen or not. However, when only cells positive for viral antigen were considered, they displayed consistently higher level of γH2Ax foci.

Furthermore, the study of Zhang et al. (2016) suggested that ZIKV-induced p53 activation might be viral strain-specific. p53 was significantly upregulated by 80.0% in ZIKV^C (Asian lineage)-infected hPSCs but did not changed in ZIKV^M (African lineage)-infected hPSCs though caspase-3 activities were dramatically activated by both ZIKV lineages (3). Treatment of hNPCs with p53 inhibitors (pifithrin-α, *p*-nitro- pifithrin-α, pifithrin-μ) dose-dependently inhibited caspase-3 activation induced by ZIKV from both clades. However, these p53 inhibitors exhibited higher potency in rescuing hNPCs from apoptosis following infection with ZIKV^C than ZIKV^M (3).

Another study by Simonin et al. (2016) also used hiPSC-derived NSCs as a model to study the differential infectivity and effect of African and

Asian lineages. The results suggested that both lineages efficiently targeted NSC; however, African strain (ArB41644) had a higher infection rate and viral production, as well as stronger antiviral response and cell death, compared to Asian strain (PF-13) (88). Gabriel et al. (2017) also evaluated the effect of virus passage number on ZIKV-induced apoptosis, and confirmed that African strain (MR766) with extensive passages had a higher level of cell apoptosis in NPCs than two low passage Asian/American strains (H/PF/2013 and FB-GWUH-2016)(89). Taken together, ZIKV-induced cell apoptosis is determined by various factors, including cell type, ZIKV lineage, and ZIKV passage history.

Garcez et al. (2016, 2017) first examined the effects of ZIKV infection in human neural stem cells (hNSCs) growing as neuropheres and brain organoids. Their results showed that ZIKV could target human brain cells, and reduced their viability and growth in neurospheres and brain organoids (4, 82). Cell cycle analyses revealed that cells in ZIKV-infected neurospheres accumulated abnormally in a sub-G1 phase, suggesting that ZIKV infection induced a reduction in cell replication associated with abnormal cell cycle progression in neurospheres (4). Furthermore, Qian et al. (2016) used a newly designed SpinΩ bioreactor-based 3D culture system to model ZIKV exposure during human brain development (83). The study confirmed that productive infection of neural progenitors with either African or Asian ZIKV led to increased cell death and reduced cell proliferation, resulting in decreased neural cell-layer volume resembling microcephaly (83).

In another study, Dang et al. (2016) used human embryonic stem cell-derived cerebral organoids to recapitulate the early stage, that is, the first trimester of fetus critical brain development (90). Their findings showed that ZIKV (MR766 strain) efficiently infected organoids and abrogated growth in cerebral organoid. It has been shown that TLR3 can be activated by ZIKV and other flaviviruses in human fibroblasts (24, 91), and that TLR3 is involved in many neuroinflammatory and neurodegenerative disorders (92–95). RT-qPCR results showed that TLR3 was up-regulated in cerebral organoids and neurospheres following ZIKV infection (90). To investigate the link between ZIKV-induced TLR activation and dysregulation of neurogenesis and apoptosis, Dang et al. (2016) evaluated the effect of both TLR3 agonist [poly(I:C)] and TLR3 inhibitor on human organoids. Neurosphere challenged with poly(I:C) exhibited a significant decrease in overall neurosphere size while treatment with TLR3 inhibitor rescued the neurosphere size reduction caused by ZIKV infection. Additional evidence showed that TLR3 competitive inhibitor attenuated the severe effects of ZIKV-induced apoptosis and organoid shrinkage seen in ZIKV-infected organoid, suggesting that TLR3 might play a pivotal role in the ZIKV-associated phenotype (90). Based on recent studies, TLR3 might play a dual role that is cell specific in which potent downstream antiviral innate immune responses are activated and tangential dysregulation of signaling networks that regulated neurogenesis and apoptosis (24, 90).

Frumence et al. (2016) used A549 cells as a model to evaluate the ZIKV-induced cell death, and demonstrated that ZIKV (strain PF-25013-18, Asian) triggered apoptosis at 48 h.p.i. by measuring the release of lactate dehydrogenace (LDH) and detecting 85 kDa-cleaved form of PARP, a hallmark of apoptosis (27). Further investigation showed that ZIKV infection led to activation of caspase-9, which plays a central role in the mitochondrial pathway of apoptosis (96). Reactive oxygen species (ROS), which are essential in driving mitochondrial apoptosis, have been shown to play a role in flavivirus-triggered apoptosis (97, 98). The results by Frumence et al. (2016) suggest that ZIKV infection stimulated the production of mitochondrial ROS in A549 cells, which might lead to the induction of apoptosis through caspase-9 activation (27).

Bowen et al. (2017) evaluated the cell viability during ZIKV infection of human moDCs, and found that contemporary and historic ZIKV isolates led to different cell fates. African lineage strains (MR766 and DakAr 41524) induced significant decrease in cell viability, while Asian lineage (P6-740 and PRVABC59) did not cause any loss of viability (48). However, the mechanism by which ZIKV infection led to cell death was not further investigated in this study. Interestingly, the results of this study seemed inconsistent with findings using the hNPC model, in which up-regulation of p53 was observed in ZIKV (Asian strain)-infected hNPCs but not in ZIKV (African strain)-infected hNPCs (81).

8.4 Induction of Autophagy by ZIKV

Autophagy or *autophagocytosis,* literally meaning "self-eating," is a highly conserved catabolic process that allows the orderly degradation and recycling of cellular components (99, 100). This process is induced under extreme conditions, such as high stress like starvation, growth factor withdrawal, viral invasion and ER stress. Under nutrient rich conditions, autophagy is inhibited via an inhibitory phosphorylation on autophagy-related protein 13 (Atg13) and Unc-51-like kinase (ULK1) by mammalian target of rapamycin complex 1 (mTORC1). Under stress conditions, this block is removed by several factors such as PTEN, AMPK, and TSC2. Activation of ULK1, which forms a complex with Atg13/FIP200/Atg101, leads to nucleation of the pre-autophagosomal structure (PAS) involving the phosphatidylino-sitol-3-kinase class III (PI3K III)-Vps34-Beclin 1 (Atg6) complex (101, 102). The subsequent elongation and completion of autophagosome is dependent on two ubiquitin-like conjugation pathways. E1-like enzyme Atg7 and E2-like enzymes Atg3 and Atg10 are involved in the conjugation of ATG12-ATG5, which forms a multimeric complex with Atg16L (Atg12-Atg5-Atg16L), while the second results in the conjugation of LC3 (Atg8) to phosphatidylethanolamine (PE) (LC3-PE). These coordinated and combined steps accomplish the formation of a mature autophagosome, which then fuses with endocytic and lysosomal compartments, ultimately leading to formation of the auto-lysosome (Figure 8.3) (103–105).

Autophagy has been implicated in the infection of several flaviviruses though its role still remains unclear (106–108). The induction of autophagy by ZIKV was first evaluated in human skin fibroblast cells (24). An electron microscopy study showed that ZIKV infection was associated with the formation of double-membrane intracytoplasmic vacuoles, which are characteristics of autophagosomes. Further functional studies indicated that autophagy might promote ZIKV replication though the exact mechanism still needs to be determined (24). The presence of autophagosomes was also observed in ZIKV-infected hNPCs by confocal microscopy by immunostaining of LC3 at the 24 h.p.i. (80). Mock-infected cells displayed a diffuse cytoplasmic pattern and low LC3 expression while ZIKV-infected cells had increased LC3 expression in the perinuclear region (24). In contrast, ZIKV infection of A549 cells did not induce the formation of autophagosomes, which was monitored using LC3 immunoblotting analysis (27). Treatment with either autophagy inhibitor 3-methyladenine (3-MA) or autophagy inducer rapamycin showed no significant change in the replication of ZIKV, indicating that autophagy might not play a role in ZIKV replication in A549 cells (27).

Liang et al. (2016) showed that ZIKV infection of human fetal NSCs (fNSCs) led to inhibition of the Akt-mTOR pathway, resulting in defective neurogenesis and aberrant activation of autophagy (84).

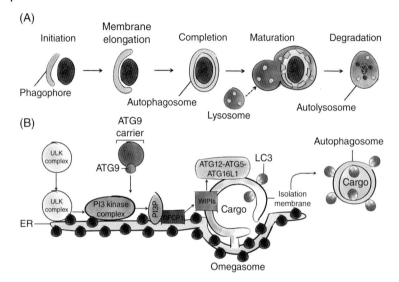

Figure 8.3 The process of autophagy and formation of autophagosome. (A) The process of autophagy. Cytoplasmic material is initially targeted by phagophore, and then membrane elongates to form an autophagosome. The autophagosome fuses with the lysosome to form an autolysosome, in which the enclosed material is degraded. (B) The molecular events during autophagosome formation. Induction of autophagy leads to the activation of the ULK1. ULK1 forms a complex with Atg13/FIP200/Atg101. The ULK complex translocates to the ER and activates PI3K complex and formation of PI3P phospholipid. The PI3P phospholipid recruits DFCP1 and WIPIs, key factors in autophagosome formation. The subsequent elongation and completion of autophagosome depends on two ubiquitin-like conjugation pathways, which leads to Atg12-Atg5-Atg16L and LC3-PE complexes. WIPIs and the ATG12-ATG5-ATG16L1 complex (outer membrane of the isolation membrane) and LC3-PE (outer and inner membrane of the isolation membrane) can emerge from subdomains of the ER. *Source:* Extracted from Mostowy et al. (2014) (133).

Many cellular signaling pathways, including PI3K-Akt-mTOR pathway, are essential for neurogenesis from NSCs, and subsequent migration and maturation (109, 110). Dysregulation of this pathway leads to several congenital brain defects including microcephaly (111, 112). By screening three structural proteins and seven nonstructural proteins of ZIKV, Liang et al. (2016) identified two viral proteins, NS4A and NS4B, which cooperatively suppressed the Akt-mTOR pathway (84). Overexpression of ZIKV NS4A and NS4B blocked neurogenesis and promoted autophagy, which benefited the ZIKV life cycle. Further functional analysis using autophagy inducer rapamycin and inhibitor 3-MA or chloroquine suggested that an efficient replication of ZIKV depended on the autophagy pathway (84).

8.5   Dysregulation of Cell Cycle and Induction of Abnormal Mitosis by ZIKV

Centrosomes are crucial in neurodevelopment not only for their key role in cell division but also for their participation in cell polarization and migration in developing brain. Severe mitotic defects have been shown to be associated with cell death and microcephaly (113, 114). Souza et al. (2016) used NPCs as a model to investigate the link between ZIKV infection and abnormalities in cell division (80). Quantification of cells with extra-numerary centrosomes showed a MOI-dependent dysregulation of centrosome replication in the ZIKV-infected culture. Additionally, five abnormalities in mitosis were observed:

1) Cell division with inadequate function of the mitotic apparatus leading to chromosome lagging and formation of micronuclei
2) Multipolar mitotic spindles causing loss of progeny
3) Mitotic catastrophe with complete loss of progeny
4) Failure in a proper cell division leading to the accumulation of aneuploidy
5) Other chromosome abnormalities, such as polyploidy, trisomy, and monosomy (80).

It is likely that ZIKV-induced reduction in cell proliferation and increase in apoptosis are partially due to the induction of mitotic dysfunction.

Defective centrosomes due to poor recruitment of centrosomal proteins are a hallmark of prematurely differentiating NPCs in microcephaly (115–117). Gabriel et al. (2017) showed that ZIKV infection indeed reduced the centrosomal recruitment of the microcephaly-linked proteins Cep164, PCNT, and CPAP (89, 118–121). In addition, serial sectioning electron microscopy analyses showed lack of appendages in mother centrioles and triplet microtubular blades at their distal ends, thus giving rise to multiple centrosomal structures during mitosis (89). This effect of ZIKV infection on centrosomal structures resembles those in cells derived from patients with microcephaly as a result of genetic defect (115, 116, 122, 123). Furthermore, ZIKV-infected organoids displayed an altered division plane of mitotic apical progenitors (115). A horizontal-oriented division plane is essential for early symmetric expansion NPCs while the change of division plane to vertical in the dividing cells is essential when neurogenesis starts after sufficient NPC expansion (115, 116, 124, 125). Compared with mock-infected organoids, ZIKV infection led to increased frequencies of vertically oriented mitotic progenitors, suggesting unscheduled switching of the division plane (89).

Onorati et al. (2016) established neuroepithelial stem (NES) cells as a model to study human neurodevelopment and ZIKV induced microcephaly (126). In this study, ZIKV (FSS 13025 strain) efficiently infected human NES cells but not NES-derived neurons. ZIKV infection caused relocation of phosphorylated TANK-binding kinase 1 (pTBK1) from mitotic centrosomes to mitochondria resulting in the disruption of mitosis and the induction of cell death (126). TBK1 is a cellular serine-threonine kinase that is essential for innate antiviral immune signaling, cell proliferation, and other processes (127–130).

Without ZIKV infection, pTBK1 was located in centrosomes in neocortical NES (NCX-NES) cells and radial glia cells (RGCs) while in the ZIKV-infected cells, pTBK1 was relocalized to mitochondria. ZIKV-infected cells also exhibited disrupted mitotic progression, including an increasing number of cells containing more than two centrosomes. As expected, TBK1 inhibitor Amlexanox impaired mitosis, provoked supernumerary centrosomes in NES cells, and exacerbated ZIKV-induced cell death (126).

References

1 Rolfe AJ, Bosco DB, Wang J, Nowakowski RS, Fan J, Ren Y. 2016. Bioinformatic analysis reveals the expression of unique transcriptomic signatures in Zika virus infected human neural stem cells. Cell Biosci **6**:42.

2 Tang H, Hammack C, Ogden SC, Wen Z, Qian X, Li Y, Yao B, Shin J, Zhang F, Lee EM, Christian KM, Didier RA, Jin P, Song H, Ming GL. 2016. Zika Virus Infects Human Cortical Neural Progenitors and Attenuates Their Growth. Cell Stem Cell **18**:587–590.

3 Zhang F, Hammack C, Ogden SC, Cheng Y, Lee EM, Wen Z, Qian X, Nguyen HN, Li Y, Yao B, Xu M, Xu T, Chen L, Wang Z, Feng H, Huang WK, Yoon KJ, Shan C, Huang L, Qin Z, Christian KM, Shi PY, Xia M, Zheng W, Wu H, Song H, Tang H, Ming GL, Jin P. 2016. Molecular signatures associated with ZIKV exposure in human cortical neural progenitors. Nucleic Acids Res **44**:8610–8620.

4 Garcez PP, Nascimento JM, de Vasconcelos JM, Madeiro da Costa R, Delvecchio R, Trindade P, Loiola EC, Higa LM, Cassoli JS, Vitoria G, Sequeira PC, Sochacki J, Aguiar RS, Fuzii HT, de Filippis AM, da Silva Goncalves Vianez Junior JL, Tanuri A, Martins-de-Souza D, Rehen SK. 2017. Zika virus disrupts molecular fingerprinting of human neurospheres. Sci Rep 7:40780.

5 Weisblum Y, Oiknine-Djian E, Vorontsov OM, Haimov-Kochman R, Zakay-Rones Z, Meir K, Shveiky D, Elgavish S, Nevo Y, Roseman M, Bronstein M, Stockheim D, From I, Eisenberg I, Lewkowicz AA, Yagel S, Panet A, Wolf DG. 2017. Zika Virus Infects Early- and Midgestation Human Maternal Decidual Tissues, Inducing Distinct Innate Tissue Responses in the Maternal-Fetal Interface. J Virol **91**.

6 Wen Z, Nguyen HN, Guo Z, Lalli MA, Wang X, Su Y, Kim NS, Yoon KJ, Shin J, Zhang C, Makri G, Nauen D, Yu H, Guzman E, Chiang CH, Yoritomo N, Kaibuchi K, Zou J, Christian KM, Cheng L, Ross CA, Margolis RL, Chen G, Kosik KS, Song H, Ming GL. 2014. Synaptic dysregulation in a human iPS cell model of mental disorders. Nature **515**:414–418.

7 Yang XJ, Ullah M. 2007. MOZ and MORF, two large MYSTic HATs in normal and cancer stem cells. Oncogene **26**:5408–5419.

8 Weisblum Y, Panet A, Haimov-Kochman R, Wolf DG. 2014. Models of vertical cytomegalovirus (CMV) transmission and pathogenesis. Semin Immunopathol 36:615–625.

9 Bayer A, Lennemann NJ, Ouyang Y, Bramley JC, Morosky S, Marques ET, Jr., Cherry S, Sadovsky Y, Coyne CB. 2016. Type III Interferons Produced by Human Placental Trophoblasts Confer Protection against Zika Virus Infection. Cell Host Microbe **19**:705–712.

10 Suthar MS, Aguirre S, Fernandez-Sesma A. 2013. Innate immune sensing of flaviviruses. PLoS Pathog **9**:e1003541.

11 Chiang JJ, Davis ME, Gack MU. 2014. Regulation of RIG-I-like receptor signaling by host and viral proteins. Cytokine Growth Factor Rev **25**:491–505.

12 Munoz-Jordan JL, Fredericksen BL. 2010. How flaviviruses activate and suppress the interferon response. Viruses **2**:676–691.

13 Ye J, Zhu B, Fu ZF, Chen H, Cao S. 2013. Immune evasion strategies of flaviviruses. Vaccine **31**:461–471.

14 Diamond MS. 2003. Evasion of innate and adaptive immunity by flaviviruses. Immunol Cell Biol **81**:196–206.

15 Loo YM, Gale M, Jr. 2011. Immune signaling by RIG-I-like receptors. Immunity **34**:680–692.

16 Kato H, Takeuchi O, Mikamo-Satoh E, Hirai R, Kawai T, Matsushita K, Hiiragi A, Dermody TS, Fujita T, Akira S. 2008. Length-dependent recognition of double-stranded ribonucleic acids by retinoic acid-inducible gene-I and melanoma differentiation-associated gene 5. J Exp Med **205**:1601–1610.

17 O'Neill LA, Golenbock D, Bowie AG. 2013. The history of Toll-like receptors—redefining innate immunity. Nat Rev Immunol **13**:453–460.

18 Uematsu S, Akira S. 2007. Toll-like receptors and Type I interferons. J Biol Chem **282**:15319–15323.

19 Diebold SS, Kaisho T, Hemmi H, Akira S, Reis e Sousa C. 2004. Innate antiviral responses by means of TLR7-mediated recognition of single-stranded RNA. Science **303**:1529–1531.

20 Lund JM, Alexopoulou L, Sato A, Karow M, Adams NC, Gale NW, Iwasaki A, Flavell RA. 2004. Recognition of single-stranded RNA viruses by Toll-like receptor 7. Proc Natl Acad Sci U S A **101**:5598–5603.

21 Wang JP, Liu P, Latz E, Golenbock DT, Finberg RW, Libraty DH. 2006. Flavivirus activation of plasmacytoid dendritic cells delineates key elements of TLR7 signaling beyond endosomal recognition. J Immunol **177**:7114–121.

22 Adachi O, Kawai T, Takeda K, Matsumoto M, Tsutsui H, Sakagami M, Nakanishi K, Akira S. 1998. Targeted disruption of the MyD88 gene results in loss of IL-1- and IL-18-mediated function. Immunity **9**:143–150.

23 Severa M, Fitzgerald KA. 2007. TLR-mediated activation of type I IFN during antiviral immune responses: fighting the battle to win the war. Curr Top Microbiol Immunol **316**:167–192.

24 Hamel R, Dejarnac O, Wichit S, Ekchariyawat P, Neyret A, Luplertlop N, Perera-Lecoin M, Surasombatpattana P, Talignani L, Thomas F, Cao-Lormeau VM, Choumet V, Briant L, Despres P, Amara A, Yssel H, Misse D. 2015. Biology of Zika Virus Infection in Human Skin Cells. J Virol **89**:8880–8896.

25 Sato M, Suemori H, Hata N, Asagiri M, Ogasawara K, Nakao K, Nakaya T, Katsuki M, Noguchi S, Tanaka N, Taniguchi T. 2000. Distinct and essential roles of transcription factors IRF-3 and IRF-7 in response to viruses for IFN-alpha/beta gene induction. Immunity **13**:539–548.

26 Lazear HM, Lancaster A, Wilkins C, Suthar MS, Huang A, Vick SC, Clepper L, Thackray L, Brassil MM, Virgin HW, Nikolich-Zugich J,

Moses AV, Gale M, Jr., Fruh K, Diamond MS. 2013. IRF-3, IRF-5, and IRF-7 coordinately regulate the type I IFN response in myeloid dendritic cells downstream of MAVS signaling. PLoS Pathog **9**:e1003118.

27 Frumence E, Roche M, Krejbich-Trotot P, El-Kalamouni C, Nativel B, Rondeau P, Misse D, Gadea G, Viranaicken W, Despres P. 2016. The South Pacific epidemic strain of Zika virus replicates efficiently in human epithelial A549 cells leading to IFN-beta production and apoptosis induction. Virology **493**:217–226.

28 Daffis S, Szretter KJ, Schriewer J, Li J, Youn S, Errett J, Lin TY, Schneller S, Zust R, Dong H, Thiel V, Sen GC, Fensterl V, Klimstra WB, Pierson TC, Buller RM, Gale M, Jr., Shi PY, Diamond MS. 2010. 2′-O methylation of the viral mRNA cap evades host restriction by IFIT family members. Nature **468**:452–456.

29 Kimura T, Katoh H, Kayama H, Saiga H, Okuyama M, Okamoto T, Umemoto E, Matsuura Y, Yamamoto M, Takeda K. 2013. Ifit1 inhibits Japanese encephalitis virus replication through binding to 5′ capped 2′-O unmethylated RNA. J Virol **87**:9997–10003.

30 Hsu YL, Shi SF, Wu WL, Ho LJ, Lai JH. 2013. Protective roles of interferon-induced protein with tetratricopeptide repeats 3 (IFIT3) in dengue virus infection of human lung epithelial cells. PLoS One **8**:e79518.

31 Savidis G, Perreira JM, Portmann JM, Meraner P, Guo Z, Green S, Brass AL. 2016. The IFITMs Inhibit Zika Virus Replication. Cell Rep **15**:2323–2330.

32 Bailey CC, Zhong G, Huang IC, Farzan M. 2014. IFITM-Family Proteins: The Cell's First Line of Antiviral Defense. Annu Rev Virol **1**:261–283.

33 Brass AL, Huang IC, Benita Y, John SP, Krishnan MN, Feeley EM, Ryan BJ, Weyer JL, van der Weyden L, Fikrig E, Adams DJ, Xavier RJ, Farzan M, Elledge SJ. 2009. The IFITM proteins mediate cellular resistance to influenza A H1N1 virus, West Nile virus, and dengue virus. Cell **139**:1243–1254.

34 Chesarino NM, McMichael TM, Hach JC, Yount JS. 2014. Phosphorylation of the antiviral protein interferon-inducible transmembrane protein 3 (IFITM3) dually regulates its endocytosis and ubiquitination. J Biol Chem **289**:11986–11992.

35 Chesarino NM, McMichael TM, Yount JS. 2014. Regulation of the trafficking and antiviral activity of IFITM3 by post-translational modifications. Future Microbiol **9**:1151–1163.

36 Chesarino NM, McMichael TM, Yount JS. 2015. E3 Ubiquitin Ligase NEDD4 Promotes Influenza Virus Infection by Decreasing Levels of the Antiviral Protein IFITM3. PLoS Pathog **11**:e1005095.

37 Everitt AR, Clare S, Pertel T, John SP, Wash RS, Smith SE, Chin CR, Feeley EM, Sims JS, Adams DJ, Wise HM, Kane L, Goulding D, Digard P, Anttila V, Baillie JK, Walsh TS, Hume DA, Palotie A, Xue Y, Colonna V, Tyler-Smith C, Dunning J, Gordon SB, Smyth RL, Openshaw PJ, Dougan G, Brass AL, Kellam P. 2012. IFITM3 restricts the morbidity and mortality associated with influenza. Nature **484**:519–523.

38 Huang Q, Chen AS, Li Q, Kang C. 2011. Expression, purification, and initial structural characterization of nonstructural protein 2B, an integral membrane protein of Dengue-2 virus, in detergent micelles. Protein Expr Purif **80**:169–175.

39 Perreira JM, Chin CR, Feeley EM, Brass AL. 2013. IFITMs restrict the replication of multiple pathogenic viruses. J Mol Biol **425**:4937–4955.

40 Tartour K, Appourchaux R, Gaillard J, Nguyen XN, Durand S, Turpin J, Beaumont E, Roch E, Berger G, Mahieux R, Brand D, Roingeard P, Cimarelli A. 2014. IFITM proteins are incorporated onto HIV-1 virion particles and negatively imprint their infectivity. Retrovirology **11**:103.

41 Feeley EM, Sims JS, John SP, Chin CR, Pertel T, Chen LM, Gaiha GD, Ryan BJ, Donis RO, Elledge SJ, Brass AL. 2011. IFITM3 inhibits influenza A virus infection by preventing cytosolic entry. PLoS Pathog **7**:e1002337.

42 Li K, Markosyan RM, Zheng YM, Golfetto O, Bungart B, Li M, Ding S, He Y, Liang C, Lee JC, Gratton E, Cohen FS, Liu SL. 2013. IFITM proteins restrict viral membrane hemifusion. PLoS Pathog **9**:e1003124.

43 Lin TY, Chin CR, Everitt AR, Clare S, Perreira JM, Savidis G, Aker AM, John SP, Sarlah D, Carreira EM, Elledge SJ, Kellam P, Brass AL. 2013. Amphotericin B increases influenza A virus infection by preventing IFITM3-mediated restriction. Cell Rep **5**:895–908.

44 Cerny D, Haniffa M, Shin A, Bigliardi P, Tan BK, Lee B, Poidinger M, Tan EY, Ginhoux F, Fink K. 2014. Selective susceptibility of human skin antigen presenting cells to productive dengue virus infection. PLoS Pathog **10**:e1004548.

45 Pinto AK, Ramos HJ, Wu X, Aggarwal S, Shrestha B, Gorman M, Kim KY, Suthar MS, Atkinson JP, Gale M, Jr., Diamond MS. 2014.

Deficient IFN signaling by myeloid cells leads to MAVS-dependent virus-induced sepsis. PLoS Pathog **10**:e1004086.

46 Robertson SJ, Lubick KJ, Freedman BA, Carmody AB, Best SM. 2014. Tick-borne flaviviruses antagonize both IRF-1 and type I IFN signaling to inhibit dendritic cell function. J Immunol **192**:2744–2755.

47 Aleyas AG, Han YW, George JA, Kim B, Kim K, Lee CK, Eo SK. 2010. Multifront assault on antigen presentation by Japanese encephalitis virus subverts CD8+ T cell responses. J Immunol **185**:1429–1441.

48 Bowen JR, Quicke KM, Maddur MS, O'Neal JT, McDonald CE, Fedorova NB, Puri V, Shabman RS, Pulendran B, Suthar MS. 2017. Zika Virus Antagonizes Type I Interferon Responses during Infection of Human Dendritic Cells. PLoS Pathog **13**:e1006164.

49 Pulendran B. 2015. The varieties of immunological experience: of pathogens, stress, and dendritic cells. Annu Rev Immunol **33**:563–606.

50 Wang JN, Ling F. 2016. Zika Virus Infection and Microcephaly: Evidence for a Causal Link. Int J Environ Res Public Health **13**.

51 Noronha L, Zanluca C, Azevedo ML, Luz KG, Santos CN. 2016. Zika virus damages the human placental barrier and presents marked fetal neurotropism. Mem Inst Oswaldo Cruz **111**:287–293.

52 Graham CH, Hawley TS, Hawley RG, MacDougall JR, Kerbel RS, Khoo N, Lala PK. 1993. Establishment and characterization of first trimester human trophoblast cells with extended lifespan. Exp Cell Res **206**:204–211.

53 Delorme-Axford E, Donker RB, Mouillet JF, Chu T, Bayer A, Ouyang Y, Wang T, Stolz DB, Sarkar SN, Morelli AE, Sadovsky Y, Coyne CB. 2013. Human placental trophoblasts confer viral resistance to recipient cells. Proc Natl Acad Sci U S A **110**:12048–12053.

54 Quicke KM, Bowen JR, Johnson EL, McDonald CE, Ma H, O'Neal JT, Rajakumar A, Wrammert J, Rimawi BH, Pulendran B, Schinazi RF, Chakraborty R, Suthar MS. 2016. Zika Virus Infects Human Placental Macrophages. Cell Host Microbe **20**:83–90.

55 Hanners NW, Eitson JL, Usui N, Richardson RB, Wexler EM, Konopka G, Schoggins JW. 2016. Western Zika Virus in Human Fetal Neural Progenitors Persists Long Term with Partial Cytopathic and Limited Immunogenic Effects. Cell Rep **15**:2315–2322.

56 Rasmussen SA, Jamieson DJ, Honein MA, Petersen LR. 2016. Zika Virus and Birth Defects--Reviewing the Evidence for Causality. N Engl J Med **374**:1981–1987.

57 Aguirre S, Maestre AM, Pagni S, Patel JR, Savage T, Gutman D, Maringer K, Bernal-Rubio D, Shabman RS, Simon V, Rodriguez-Madoz JR, Mulder LC, Barber GN, Fernandez-Sesma A. 2012. DENV inhibits type I IFN production in infected cells by cleaving human STING. PLoS Pathog **8**:e1002934.

58 Liu WJ, Chen HB, Wang XJ, Huang H, Khromykh AA. 2004. Analysis of adaptive mutations in Kunjin virus replicon RNA reveals a novel role for the flavivirus nonstructural protein NS2A in inhibition of beta interferon promoter-driven transcription. J Virol **78**:12225–12235.

59 Liu WJ, Wang XJ, Mokhonov VV, Shi PY, Randall R, Khromykh AA. 2005. Inhibition of interferon signaling by the New York 99 strain and Kunjin subtype of West Nile virus involves blockage of STAT1 and STAT2 activation by nonstructural proteins. J Virol **79**:1934–1942.

60 Munoz-Jordan JL, Sanchez-Burgos GG, Laurent-Rolle M, Garcia-Sastre A. 2003. Inhibition of interferon signaling by dengue virus. Proc Natl Acad Sci U S A **100**:14333–14338.

61 Ashour J, Laurent-Rolle M, Shi PY, Garcia-Sastre A. 2009. NS5 of dengue virus mediates STAT2 binding and degradation. J Virol **83**:5408–5418.

62 Best SM, Morris KL, Shannon JG, Robertson SJ, Mitzel DN, Park GS, Boer E, Wolfinbarger JB, Bloom ME. 2005. Inhibition of interferon-stimulated JAK-STAT signaling by a tick-borne flavivirus and identification of NS5 as an interferon antagonist. J Virol **79**:12828–12839.

63 Laurent-Rolle M, Morrison J, Rajsbaum R, Macleod JM, Pisanelli G, Pham A, Ayllon J, Miorin L, Martinez-Romero C, tenOever BR, Garcia-Sastre A. 2014. The interferon signaling antagonist function of yellow fever virus NS5 protein is activated by type I interferon. Cell Host Microbe **16**:314–327.

64 Laurent-Rolle M, Boer EF, Lubick KJ, Wolfinbarger JB, Carmody AB, Rockx B, Liu W, Ashour J, Shupert WL, Holbrook MR, Barrett AD, Mason PW, Bloom ME, Garcia-Sastre A, Khromykh AA, Best SM. 2010. The NS5 protein of the virulent West Nile virus NY99 strain is a potent antagonist of type I interferon-mediated JAK-STAT signaling. J Virol **84**:3503–3515.

65 Lin RJ, Chang BL, Yu HP, Liao CL, Lin YL. 2006. Blocking of interferon-induced Jak-Stat signaling by Japanese encephalitis virus NS5 through a protein tyrosine phosphatase-mediated mechanism. J Virol **80**:5908–5918.

66 Lubick KJ, Robertson SJ, McNally KL, Freedman BA, Rasmussen AL, Taylor RT, Walts AD, Tsuruda S, Sakai M, Ishizuka M, Boer EF, Foster EC, Chiramel AI, Addison CB, Green R, Kastner DL, Katze MG, Holland SM, Forlino A, Freeman AF, Boehm M, Yoshii K, Best SM. 2015. Flavivirus Antagonism of Type I Interferon Signaling Reveals Prolidase as a Regulator of IFNAR1 Surface Expression. Cell Host Microbe **18**:61–74.

67 Mazzon M, Jones M, Davidson A, Chain B, Jacobs M. 2009. Dengue virus NS5 inhibits interferon-alpha signaling by blocking signal transducer and activator of transcription 2 phosphorylation. J Infect Dis **200**:1261–1270.

68 Morrison J, Laurent-Rolle M, Maestre AM, Rajsbaum R, Pisanelli G, Simon V, Mulder LC, Fernandez-Sesma A, Garcia-Sastre A. 2013. Dengue virus co-opts UBR4 to degrade STAT2 and antagonize type I interferon signaling. PLoS Pathog **9**:e1003265.

69 Grant A, Ponia SS, Tripathi S, Balasubramaniam V, Miorin L, Sourisseau M, Schwarz MC, Sanchez-Seco MP, Evans MJ, Best SM, Garcia-Sastre A. 2016. Zika Virus Targets Human STAT2 to Inhibit Type I Interferon Signaling. Cell Host Microbe **19**:882–890.

70 Ghosh Roy S, Sadigh B, Datan E, Lockshin RA, Zakeri Z. 2014. Regulation of cell survival and death during Flavivirus infections. World J Biol Chem **5**:93–105.

71 McLean JE, Ruck A, Shirazian A, Pooyaei-Mehr F, Zakeri ZF. 2008. Viral manipulation of cell death. Curr Pharm Des **14**:198–220.

72 Clarke PG, Clarke S. 1996. Nineteenth century research on naturally occurring cell death and related phenomena. Anat Embryol (Berl) **193**:81–99.

73 Proskuryakov SY, Konoplyannikov AG, Gabai VL. 2003. Necrosis: a specific form of programmed cell death? Exp Cell Res **283**:1–16.

74 Ouyang L, Shi Z, Zhao S, Wang FT, Zhou TT, Liu B, Bao JK. 2012. Programmed cell death pathways in cancer: a review of apoptosis, autophagy and programmed necrosis. Cell Prolif **45**:487–498.

75 Thomas LR, Johnson RL, Reed JC, Thorburn A. 2004. The C-terminal tails of tumor necrosis factor-related apoptosis-inducing ligand (TRAIL) and Fas receptors have opposing functions in Fas-associated death domain (FADD) recruitment and can regulate agonist-specific mechanisms of receptor activation. J Biol Chem **279**:52479–52486.

76 Barnhart BC, Lee JC, Alappat EC, Peter ME. 2003. The death effector domain protein family. Oncogene **22**:8634–8644.

77 Dempsey PW, Doyle SE, He JQ, Cheng G. 2003. The signaling adaptors and pathways activated by TNF superfamily. Cytokine Growth Factor Rev **14**:193–209.

78 Kerr JF, Wyllie AH, Currie AR. 1972. Apoptosis: a basic biological phenomenon with wide-ranging implications in tissue kinetics. Br J Cancer **26**:239–257.

79 Ghobrial IM, Witzig TE, Adjei AA. 2005. Targeting apoptosis pathways in cancer therapy. CA Cancer J Clin **55**:178–194.

80 Souza BS, Sampaio GL, Pereira CS, Campos GS, Sardi SI, Freitas LA, Figueira CP, Paredes BD, Nonaka CK, Azevedo CM, Rocha VP, Bandeira AC, Mendez-Otero R, Dos Santos RR, Soares MB. 2016. Zika virus infection induces mitosis abnormalities and apoptotic cell death of human neural progenitor cells. Sci Rep **6**:39775.

81 Ghouzzi VE, Bianchi FT, Molineris I, Mounce BC, Berto GE, Rak M, Lebon S, Aubry L, Tocco C, Gai M, Chiotto AM, Sgro F, Pallavicini G, Simon-Loriere E, Passemard S, Vignuzzi M, Gressens P, Di Cunto F. 2016. ZIKA virus elicits P53 activation and genotoxic stress in human neural progenitors similar to mutations involved in severe forms of genetic microcephaly and p53. Cell Death Dis **7**:e2440.

82 Garcez PP, Loiola EC, Madeiro da Costa R, Higa LM, Trindade P, Delvecchio R, Nascimento JM, Brindeiro R, Tanuri A, Rehen SK. 2016. Zika virus impairs growth in human neurospheres and brain organoids. Science **352**:816–818.

83 Qian X, Nguyen HN, Song MM, Hadiono C, Ogden SC, Hammack C, Yao B, Hamersky GR, Jacob F, Zhong C, Yoon KJ, Jeang W, Lin L, Li Y, Thakor J, Berg DA, Zhang C, Kang E, Chickering M, Nauen D, Ho CY, Wen Z, Christian KM, Shi PY, Maher BJ, Wu H, Jin P, Tang H, Song H, Ming GL. 2016. Brain-Region-Specific Organoids Using Mini-bioreactors for Modeling ZIKV Exposure. Cell **165**:1238–1254.

84 Liang Q, Luo Z, Zeng J, Chen W, Foo SS, Lee SA, Ge J, Wang S, Goldman SA, Zlokovic BV, Zhao Z, Jung JU. 2016. Zika Virus NS4A and NS4B Proteins Deregulate Akt-mTOR Signaling in Human Fetal Neural Stem Cells to Inhibit Neurogenesis and Induce Autophagy. Cell Stem Cell **19**:663–671.

85 Ryan EL, Hollingworth R, Grand RJ. 2016. Activation of the DNA Damage Response by RNA Viruses. Biomolecules **6**:2.

86 Luftig MA. 2014. Viruses and the DNA Damage Response: Activation and Antagonism. Annu Rev Virol **1**:605–625.

87 Turinetto V, Giachino C. 2015. Multiple facets of histone variant H2AX: a DNA double-strand-break marker with several biological functions. Nucleic Acids Res **43**:2489–2498.

88 Simonin Y, Loustalot F, Desmetz C, Foulongne V, Constant O,
Fournier-Wirth C, Leon F, Moles JP, Goubaud A, Lemaitre JM,
Maquart M, Leparc-Goffart I, Briant L, Nagot N, Van de Perre P,
Salinas S. 2016. Zika Virus Strains Potentially Display Different
Infectious Profiles in Human Neural Cells. EBioMedicine **12**:161–169.

89 Gabriel E, Ramani A, Karow U, Gottardo M, Natarajan K, Gooi LM,
Goranci-Buzhala G, Krut O, Peters F, Nikolic M, Kuivanen S,
Korhonen E, Smura T, Vapalahti O, Papantonis A, Schmidt-Chanasit
J, Riparbelli M, Callaini G, Kronke M, Utermohlen O,
Gopalakrishnan J. 2017. Recent Zika Virus Isolates Induce Premature
Differentiation of Neural Progenitors in Human Brain Organoids.
Cell Stem Cell **20**:397–406 e395.

90 Dang J, Tiwari SK, Lichinchi G, Qin Y, Patil VS, Eroshkin AM, Rana
TM. 2016. Zika Virus Depletes Neural Progenitors in Human
Cerebral Organoids through Activation of the Innate Immune
Receptor TLR3. Cell Stem Cell **19**:258–265.

91 Tsai YT, Chang SY, Lee CN, Kao CL. 2009. Human TLR3 recognizes
dengue virus and modulates viral replication in vitro. Cell Microbiol
11:604–615.

92 Cameron JS, Alexopoulou L, Sloane JA, DiBernardo AB, Ma Y,
Kosaras B, Flavell R, Strittmatter SM, Volpe J, Sidman R, Vartanian T.
2007. Toll-like receptor 3 is a potent negative regulator of axonal
growth in mammals. J Neurosci **27**:13033–13041.

93 Lathia JD, Okun E, Tang SC, Griffioen K, Cheng A, Mughal MR,
Laryea G, Selvaraj PK, ffrench-Constant C, Magnus T, Arumugam
TV, Mattson MP. 2008. Toll-like receptor 3 is a negative regulator of
embryonic neural progenitor cell proliferation. J Neurosci
28:13978–13984.

94 Okun E, Griffioen K, Barak B, Roberts NJ, Castro K, Pita MA, Cheng
A, Mughal MR, Wan R, Ashery U, Mattson MP. 2010. Toll-like
receptor 3 inhibits memory retention and constrains adult
hippocampal neurogenesis. Proc Natl Acad Sci U S A
107:15625–15630.

95 Okun E, Griffioen KJ, Mattson MP. 2011. Toll-like receptor signaling
in neural plasticity and disease. Trends Neurosci **34**:269–281.

96 Susin SA, Lorenzo HK, Zamzami N, Marzo I, Brenner C, Larochette
N, Prevost MC, Alzari PM, Kroemer G. 1999. Mitochondrial release
of caspase-2 and -9 during the apoptotic process. J Exp Med
189:381–394.

97 Fleury C, Mignotte B, Vayssiere JL. 2002. Mitochondrial reactive
oxygen species in cell death signaling. Biochimie **84**:131–141.

98 Olagnier D, Peri S, Steel C, van Montfoort N, Chiang C, Beljanski V, Slifker M, He Z, Nichols CN, Lin R, Balachandran S, Hiscott J. 2014. Cellular oxidative stress response controls the antiviral and apoptotic programs in dengue virus-infected dendritic cells. PLoS Pathog **10**:e1004566.

99 Mizushima N, Komatsu M. 2011. Autophagy: renovation of cells and tissues. Cell **147**:728–741.

100 Kobayashi S. 2015. Choose Delicately and Reuse Adequately: The Newly Revealed Process of Autophagy. Biol Pharm Bull **38**:1098–1103.

101 Lee J, Yang KH, Joe CO, Kang SS. 2011. Formation of distinct inclusion bodies by inhibition of ubiquitin-proteasome and autophagy-lysosome pathways. Biochem Biophys Res Commun **404**:672–677.

102 Kaur J, Debnath J. 2015. Autophagy at the crossroads of catabolism and anabolism. Nat Rev Mol Cell Biol **16**:461–472.

103 Kuma A, Mizushima N, Ishihara N, Ohsumi Y. 2002. Formation of the approximately 350-kDa Apg12-Apg5.Apg16 multimeric complex, mediated by Apg16 oligomerization, is essential for autophagy in yeast. J Biol Chem **277**:18619–18625.

104 Tanida I, Ueno T, Kominami E. 2004. LC3 conjugation system in mammalian autophagy. Int J Biochem Cell Biol **36**:2503–2518.

105 Geng J, Klionsky DJ. 2008. The Atg8 and Atg12 ubiquitin-like conjugation systems in macroautophagy. 'Protein modifications: beyond the usual suspects' review series. EMBO Rep **9**:859–864.

106 Heaton NS, Randall G. 2011. Dengue virus and autophagy. Viruses **3**:1332–1341.

107 Lee YR, Hu HY, Kuo SH, Lei HY, Lin YS, Yeh TM, Liu CC, Liu HS. 2013. Dengue virus infection induces autophagy: an in vivo study. J Biomed Sci **20**:65.

108 Martin-Acebes MA, Blazquez AB, Saiz JC. 2015. Reconciling West Nile virus with the autophagic pathway. Autophagy **11**:861–864.

109 Lee DY. 2015. Roles of mTOR Signaling in Brain Development. Exp Neurobiol **24**:177–185.

110 Wahane SD, Hellbach N, Prentzell MT, Weise SC, Vezzali R, Kreutz C, Timmer J, Krieglstein K, Thedieck K, Vogel T. 2014. PI3K-p110-alpha-subtype signalling mediates survival, proliferation and neurogenesis of cortical progenitor cells via activation of mTORC2. J Neurochem **130**:255–267.

111 Cloetta D, Thomanetz V, Baranek C, Lustenberger RM, Lin S, Oliveri F, Atanasoski S, Ruegg MA. 2013. Inactivation of mTORC1 in the developing brain causes microcephaly and affects gliogenesis. J Neurosci **33**:7799–7810.

112 Mirzaa GM, Riviere JB, Dobyns WB. 2013. Megalencephaly syndromes and activating mutations in the PI3K-AKT pathway: MPPH and MCAP. Am J Med Genet C Semin Med Genet **163C**:122–130.

113 Marthiens V, Rujano MA, Pennetier C, Tessier S, Paul-Gilloteaux P, Basto R. 2013. Centrosome amplification causes microcephaly. Nat Cell Biol **15**:731–740.

114 Thornton GK, Woods CG. 2009. Primary microcephaly: do all roads lead to Rome? Trends Genet **25**:501–510.

115 Gabriel E, Wason A, Ramani A, Gooi LM, Keller P, Pozniakovsky A, Poser I, Noack F, Telugu NS, Calegari F, Saric T, Hescheler J, Hyman AA, Gottardo M, Callaini G, Alkuraya FS, Gopalakrishnan J. 2016. CPAP promotes timely cilium disassembly to maintain neural progenitor pool. EMBO J **35**:803–819.

116 Lancaster MA, Renner M, Martin CA, Wenzel D, Bicknell LS, Hurles ME, Homfray T, Penninger JM, Jackson AP, Knoblich JA. 2013. Cerebral organoids model human brain development and microcephaly. Nature **501**:373–379.

117 Zheng X, Gooi LM, Wason A, Gabriel E, Mehrjardi NZ, Yang Q, Zhang X, Debec A, Basiri ML, Avidor-Reiss T, Pozniakovsky A, Poser I, Saric T, Hyman AA, Li H, Gopalakrishnan J. 2014. Conserved TCP domain of Sas-4/CPAP is essential for pericentriolar material tethering during centrosome biogenesis. Proc Natl Acad Sci U S A **111**:E354–363.

118 Bond J, Roberts E, Springell K, Lizarraga SB, Scott S, Higgins J, Hampshire DJ, Morrison EE, Leal GF, Silva EO, Costa SM, Baralle D, Raponi M, Karbani G, Rashid Y, Jafri H, Bennett C, Corry P, Walsh CA, Woods CG. 2005. A centrosomal mechanism involving CDK5RAP2 and CENPJ controls brain size. Nat Genet **37**:353–355.

119 Cizmecioglu O, Arnold M, Bahtz R, Settele F, Ehret L, Haselmann-Weiss U, Antony C, Hoffmann I. 2010. Cep152 acts as a scaffold for recruitment of Plk4 and CPAP to the centrosome. J Cell Biol **191**:731–739.

120 Guernsey DL, Jiang H, Hussin J, Arnold M, Bouyakdan K, Perry S, Babineau-Sturk T, Beis J, Dumas N, Evans SC, Ferguson M, Matsuoka M, Macgillivray C, Nightingale M, Patry L, Rideout AL,

Thomas A, Orr A, Hoffmann I, Michaud JL, Awadalla P, Meek DC, Ludman M, Samuels ME. 2010. Mutations in centrosomal protein CEP152 in primary microcephaly families linked to MCPH4. Am J Hum Genet **87**:40–51.

121 Rauch A, Thiel CT, Schindler D, Wick U, Crow YJ, Ekici AB, van Essen AJ, Goecke TO, Al-Gazali L, Chrzanowska KH, Zweier C, Brunner HG, Becker K, Curry CJ, Dallapiccola B, Devriendt K, Dorfler A, Kinning E, Megarbane A, Meinecke P, Semple RK, Spranger S, Toutain A, Trembath RC, Voss E, Wilson L, Hennekam R, de Zegher F, Dorr HG, Reis A. 2008. Mutations in the pericentrin (PCNT) gene cause primordial dwarfism. Science **319**:816–819.

122 Hussain MS, Baig SM, Neumann S, Nurnberg G, Farooq M, Ahmad I, Alef T, Hennies HC, Technau M, Altmuller J, Frommolt P, Thiele H, Noegel AA, Nurnberg P. 2012. A truncating mutation of CEP135 causes primary microcephaly and disturbed centrosomal function. Am J Hum Genet **90**:871–878.

123 Kalay E, Yigit G, Aslan Y, Brown KE, Pohl E, Bicknell LS, Kayserili H, Li Y, Tuysuz B, Nurnberg G, Kiess W, Koegl M, Baessmann I, Buruk K, Toraman B, Kayipmaz S, Kul S, Ikbal M, Turner DJ, Taylor MS, Aerts J, Scott C, Milstein K, Dollfus H, Wieczorek D, Brunner HG, Hurles M, Jackson AP, Rauch A, Nurnberg P, Karaguzel A, Wollnik B. 2011. CEP152 is a genome maintenance protein disrupted in Seckel syndrome. Nat Genet **43**:23–26.

124 Rakic P. 1995. A small step for the cell, a giant leap for mankind: a hypothesis of neocortical expansion during evolution. Trends Neurosci **18**:383–388.

125 Yingling J, Youn YH, Darling D, Toyo-Oka K, Pramparo T, Hirotsune S, Wynshaw-Boris A. 2008. Neuroepithelial stem cell proliferation requires LIS1 for precise spindle orientation and symmetric division. Cell **132**:474–486.

126 Onorati M, Li Z, Liu F, Sousa AM, Nakagawa N, Li M, Dell'Anno MT, Gulden FO, Pochareddy S, Tebbenkamp AT, Han W, Pletikos M, Gao T, Zhu Y, Bichsel C, Varela L, Szigeti-Buck K, Lisgo S, Zhang Y, Testen A, Gao XB, Mlakar J, Popovic M, Flamand M, Strittmatter SM, Kaczmarek LK, Anton ES, Horvath TL, Lindenbach BD, Sestan N. 2016. Zika Virus Disrupts Phospho-TBK1 Localization and Mitosis in Human Neuroepithelial Stem Cells and Radial Glia. Cell Rep **16**:2576–2592.

127 Farlik M, Rapp B, Marie I, Levy DE, Jamieson AM, Decker T. 2012. Contribution of a TANK-binding kinase 1-interferon (IFN)

regulatory factor 7 pathway to IFN-gamma-induced gene expression. Mol Cell Biol **32**:1032–1043.

128 Helgason E, Phung QT, Dueber EC. 2013. Recent insights into the complexity of Tank-binding kinase 1 signaling networks: the emerging role of cellular localization in the activation and substrate specificity of TBK1. FEBS Lett **587**:1230–1237.

129 Pillai S, Nguyen J, Johnson J, Haura E, Coppola D, Chellappan S. 2015. Tank binding kinase 1 is a centrosome-associated kinase necessary for microtubule dynamics and mitosis. Nat Commun **6**:10072.

130 Thurston TL, Ryzhakov G, Bloor S, von Muhlinen N, Randow F. 2009. The TBK1 adaptor and autophagy receptor NDP52 restricts the proliferation of ubiquitin-coated bacteria. Nat Immunol **10**:1215–1221.

131 Furr SR, Marriott I. 2012. Viral CNS infections: role of glial pattern recognition receptors in neuroinflammation. Front Microbiol **3**:201.

132 Galluzzi L, Brenner C, Morselli E, Touat Z, Kroemer G. 2008. Viral control of mitochondrial apoptosis. PLoS Pathog **4**:e1000018.

133 Mostowy S. 2014. Multiple roles of the cytoskeleton in bacterial autophagy. PLoS Pathog **10**:e1004409.

9

Inhibitors of ZIKV Replication and Infection

Despite its initial identification in 1947, ZIKV has only recently emerged as an important infectious agent because of its association with microcephaly and neuronal problems in fetuses. There are ongoing intensive efforts to develop a vaccine or an effective drug to limit its pathogenesis and impact on the human health and society. As it is now, there is no vaccine available for the treatment and prevention of ZIKV infection. There is also no Food and Drug Administration (FDA)-approved drug against ZIKV. This chapter reviews and summarizes the most recent developments of drugs that inhibit ZIKV infection and replication, as well as ZIKV-related pathogenesis.

9.1 Drugs That Lead to the Destruction of ZIKV Virions

Epigallocatechin gallate (EGCG) is a polyphenol present in large quantities in green tea (Figure 9.1). It has been shown that EGCG has a general, intense antiviral activity against many viruses, including the human immunodeficiency virus (HIV), herpes simplex virus (HSV), influenza virus, hepatitis B virus, and hepatitis C virus (1–6). The main mechanism of action of EGCG is inhibition of virus entry into host cells.

Carneiro et al. (2016) recently reported that pre-incubation of ZIKV with EGCG potently inhibited the replication of the Brazilian strain (ZIKV^{BR}) and MR766 strain with EC$_{50}$ of approximately 21.4 μM (7). EGCG caused 85.0% inhibition of MR766 strain at 25 μM whereas

Zika Virus and Diseases: From Molecular Biology to Epidemiology, First Edition.
Suzane Ramos da Silva, Fan Cheng and Shou-Jiang Gao.
© 2018 John Wiley & Sons, Inc. Published 2018 by John Wiley & Sons, Inc.

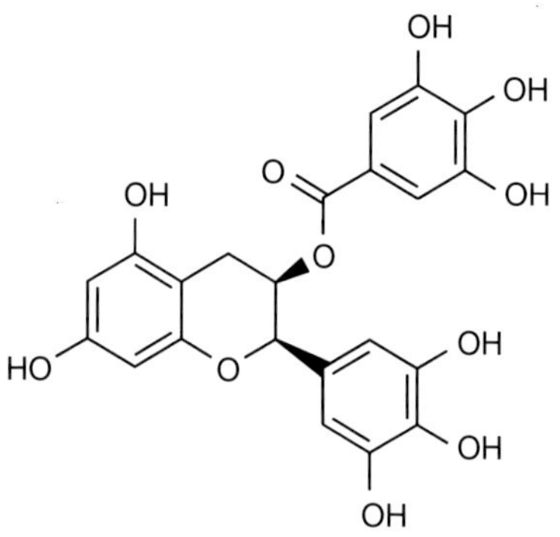

Figure 9.1 Structural formula of epigallocatechin gallate (EGCG). *Source:* Extracted from Carneiro et al. (2016) (7).

$50\,\mu M$ are required to achieve the same inhibition effect on ZIKV[BR]. It was hypothesized that EGCG exerted a direct action on virus particle through destroying the envelope phospholipids as demonstrated in the HIV infection, thus leading to the destruction of the virion (8). However, further investigation is needed to test this hypothesis.

9.2 Drugs That Inhibit ZIKV Entry and Endocytosis

9.2.1 Chloroquine and Derivatives

Chloroquine, a 4-aminoquinoline, is a weak base that can be rapidly imported into acidic vesicles, thus increasing their pH (Figure 9.2) (9). It is a medication approved by the FDA to prevent and treat malaria. Chloroquine is also a nonspecific antiviral agent against a large spectrum of viruses *in vitro*. Through the inhibition of pH-dependent steps of viral replication, chloroquine restricts the infection of HIV, influenza, DENV, JEV, and WNV (10–14).

A study by Delvecchio et al. (2016) showed that chloroquine exhibited antiviral activity against ZIKV in Vero cells, human brain microvascular endothelial cells, human neural stem cells, and mouse neurospheres (15). Chloroquine inhibited ZIKV infection of cells. The EC50, the concentration at which chloroquine protects 50% of the cells from ZIKV infection, was 9.82 to $14.2\,\mu M$ depending on the cell type; while the CC50, the 50% cytotoxicity concentration, ranged from 94.95 to $134.54\,\mu M$. When the values of EC50 were compared among different virus types, the EC50 of chloroquine for ZIKV MR766 is much lower than those for DENV ($25\,\mu M$) and HIV ($100\,\mu M$) (10, 12, 15).

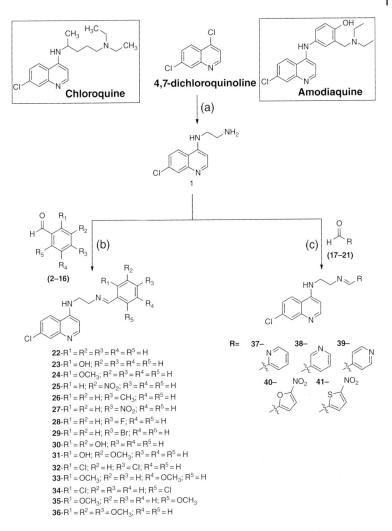

Figure 9.2 Synthesis of N-(2-(arylmethylimino)ethyl)-7-chloroquinolin-4-amine derivatives. *Source:* Reproduced from G. Barbosa-Lima et al. N-(2-(arylmethylimino)ethyl)-7-chloroquinolin-4-amine derivatives, synthesized by thermal and ultrasonic means, are endowed with anti-Zika virus activity. *European Journal of Medicinal Chemistry.* 2017; 127:434–441. Copyright © 2017 Elsevier Masson SAS. All rights reserved (18).

Two other features of chloroquine render it a potential therapeutic agent for treating ZIKV infection:

1) Chloroquine is an FDA approved medication for treating malaria; its usage during pregnancy has been thoroughly evaluated. No increment in birth defects was observed when prophylactic doses of chloroquine (400 mg/week) were administered for malaria (16).
2) Chloroquine, as well as its analogue hydroxychloroquine, is widely distributed to body tissues. Chloroquine is capable of penetrating the placenta barrier and reaching similar concentrations in both maternal and fetal plasma (17).

A recent study by Barbosa-Lima et al. (2017) further demonstrated that modifications to the chloroquine structure might lead to more effective anti-ZIKV agents (18). A series of N-(2-(arylmethylimino) ethyl)-7-chloroquinolin-4-amine derivatives inhibit ZIKV replication *in vitro* (Figure 9.2). The quinolone derivative, N-(2-((5-nitrofuran-2-yl)methylimino)ethyl)-7-chloroquinolin-4-amine, compound 40, is the most potent compound in the series, exhibiting an EC50 value of $0.8 \pm 0.07\,\mu M$, compared to that of chloroquine of $12 \pm 3.2\,\mu M$ (18). This series of compounds could be a good follow-up point for new anti-ZIKV drugs.

9.2.2 Mefloquine and Derivatives

Mefloquine, under the brand names Lariam, is a medication approved by the FDA for prevention and treatment of malaria (Figure 9.3) (19). Mefloquine also has anticancer, antituberculosis, and antiviral activities (20–26). Due to the high lipophilicity of mefloquine, it can easily cross the blood-brain barrier and reach the central nervous system (27–30), making it a lead molecule for treating neurotropic viruses (20).

Mefloquine was first demonstrated to be able to inhibit ZIKV replication in a screen of 774 FDA-approved drugs for anti-ZIKV inhibitors using human hepatoma cell line (31). In this report, Barrows et al. (2016) revealed the effect of mefloquine on ZIKV (MEX_I_7/2015 strain) infection in different cell types. In HeLa cells, mefloquine at $10\,\mu M$ significantly reduced the virus infection rate by over 90.0% inhibition but showed no effect at $1\,\mu M$. In human neural stem cells, mefloquine at $1\,\mu M$ cannot inhibit the infection ZIKV of both MEX_I_7/2015 and DAK_41525 (African) strains (31). In primary human amnion

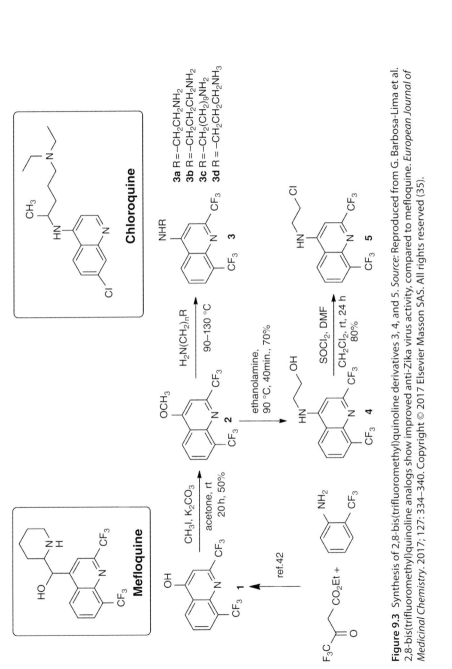

Figure 9.3 Synthesis of 2,8-bis(trifluoromethyl)quinoline derivatives 3, 4, and 5. *Source:* Reproduced from G. Barbosa-Lima et al. 2,8-bis(trifluoromethyl)quinoline analogs show improved anti-Zika virus activity, compared to mefloquine. *European Journal of Medicinal Chemistry.* 2017; 127: 334–340. Copyright © 2017 Elsevier Masson SAS. All rights reserved (35).

epithelial cells (HAECs), mefloquine at 16 μM exhibited strong inhibition of ZIKV MEX_I_7/2015 infectivity without causing substantial cell toxicity (31).

Mefloquine has been shown to cross the placenta (32), and can be used to treat both the mother and the fetus. Mefloquine is an FDA-approved pregnancy category B drug, indicating that controlled studies in pregnant women have not shown an increased risk of fetal abnormalities, although some adverse findings have occurred in animals. On the other hand, it indicates that there is no adequate human study. However, animal studies show no fetal risk, albeit there is a remote possibility that it might have side effects on the fetus. The use of mefloquine to treat ZIKV infection during pregnancy might require a shorter course and gestational age windows that are different from the currently accepted indications for better safety profile.

In the aforementioned study, mefloquine inhibited ZIKV infection *in vitro* at 10 μM, which is approximately 2.2-fold higher than its potential maximum serum concentration (Cmax) (1,872 ng/L) (33). In order to improve the potency of mefloquine, Barbosa-Lima et al. (2017) used medicinal chemistry-driven approaches to synthesize a series of new 2,8-bis(trifluoromethyl)quinoline derivatives (Figure 9.3) and evaluated their ability to inhibit ZIKV replication *in vitro* (34). Most of the derivatives in the series showed antiviral activities similar to that of mefloquine. Quinoline derivative 3a and 4 are the most potent compounds, both with mean EC50 values of 0.8 μM, which is approximately five times lower than that of mefloquine. Compounds 3b and 5 showed approximately two and three times higher potency than the reference molecule (34).

In vivo experimental models for ZIKV infection has revealed virus tropism for peripheral organs, including the spleen, liver, kidney, and testis before it reaches the central nervous system (35–37). Considering the higher permeability of these organs to lipophilic compounds, mefloquine is a more attractive anti-ZIKV compound than chloroquine. Modification of the C4-position of the melfoquine nucleus further enhanced its antiviral activity. These derivatives are promising candidates as novel anti-ZIKV drugs (34).

9.2.3 Obatoclax and Saliphenylhalamide (SaliPhe)

Obatoclax and SaliPhe are two cell-directed anticancer compounds, which showed broad-spectrum antiviral activity against several human viruses in human cell lines (38–41). Obatoclax, an inhibitor of the Bcl-2 family of proteins (42), is able to inhibit endocytosis of influenza A and B viruses (FLUAV and FLUBV), Bunyamvera virus (BUNV), and Sindbis virus (SINV). SaliPhe is a potent inhibitor of vacuolar-ATPase activity, preventing the acidification of endosomes. It arrests FLUAV, FLUBV, BUNV, SINV, and papilloma virus in the endocytic compartments.

In the latest study by Kuivanen et al. (2017), the anti-ZIKV activities of these two compounds were evaluated using human retinal pigment epithelial (RPE) cells, which represent natural targets for ZIKV infection (43, 44). The ZIKV strain used in this study is FB-GWUH-2016 strain, which efficiently kills the RPE cells and represents a clearly disease-associated, low-passage strain isolated from a clinical case of severe congenital infection (45). Using a cell viability/drug rescue assay, both obatoclax and SaliPhe were shown to rescue infected cells from ZIKV-induced cell death in a dose-dependent manner. At the noncytotoxic concentrations, both compounds prevented the synthesis of viral proteins at 24 hours post-infection (h.p.i.), and the synthesis of viral genomic RNA and production of progeny virus at 48 h.p.i. In addition, cells were efficiently protected from virus-induced death when compounds were applied before 4 h.p.i., indicating that these two compounds exert antiviral activity by targeting the early stage of ZIKV infection, most likely by inhibiting the endocytic uptake of ZIKV (44).

9.2.4   Niclosamide

Niclosamide, under the trade name Niclocide, is an FDA-approved drug for treating worm infections in both humans and livestock (46–48). It has been shown to inhibit several viruses *in vitro*, including SARS coronavirus, and JEV (49–51). Its broad antiviral activity has been attributed to its ability to neutralize endosomal pH and interfere with acidic pH-dependent membrane fusion (50), an essential step for infection of a large variety of viruses. A study by Xu et al. (2016) showed that niclosamide inhibited the replication of ZIKV strains MR766 (1947 Ugandan strain), FSS13025 (2010 Cambodian strain), and PRVABC59 (2015 Puerto Rico strain) in a dose-dependent manner, as measured by NS1 expression in glioblastoma SNB-19 cells (52). Further measurement of intracellular ZIKV genomic RNA levels showed an IC50 value of 0.37 µM. Niclosamide suppressed the production of progeny ZIKV virions at submicromolar concentrations. The time-of-addition experiment showed that niclosamide inhibited ZIKV infection when added at either 1 hour before or 4 hours after infection. Time-course analysis showed that the reduction of viral RNA by niclosamide was apparent when it was used at the entry phase (0–4 h.p.i.), which correlated with the replication phase (4–24 h.p.i.) of infection cycle, indicating that niclosamide inhibited ZIKV infection at a post-entry stage probably during viral RNA replication step (52).

As an FDA-approved drug, niclosamide is well tolerated in humans. Niclosamide is a pregnancy category B drug, indicating that no risk to fetuses in animal studies. However, "the risk of treatment (with Niclosamide) in pregnant women who are known to have an infection needs to be balanced with the risk of disease progression in the absence of treatment." It has low toxicity in mammals with an oral median lethal dose (LD50) value of 5,000 mg per kg body weight (mg/kg) in rats (53). The potency of niclosamide on inhibition of ZIKV replication is in the submicromolar range, whereas clinically it can be delivered at micromolar levels (53).

9.3 Drugs That Target ZIKV NS2B-NS3 Protease Activity

ZIKV NS2B-NS3 protease is an attractive antiviral target due to its essential role in processing viral polypeptide to generate structural and nonstructural proteins during viral replication.

9.3.1 Bromocriptine

Bromocriptine (brand name: Parlodel, Cycloset), an ergoline derivative, is an agonist of dopamine receptors 2 and 3, which is used to treat galactorrhea and Parkinson's disease (54–56). Bromocriptine has been shown to have antiviral activity against other flaviviruses, including DENV serotypes 1-4 and tick-borne encephalitis virus (57). In a drug repurposing study, Chan et al. (2017) found that bromocriptine had inhibitory effects on ZIKV replication in cytopathic effect inhibition, virus yield reduction, and plaque reduction assays (58). The time-of-drug-addition assay showed that bromocriptine exerted antiviral activity between 0 and 12 h.p.i., corroborating with post-entry events in the viral replication cycle prior to budding. In the docking model, bromocriptine was predicted to interact with active site residues of the proteolytic cavity involving His51 and Ser135. Furthermore, a fluorescence-based protease function assay confirmed that bromocriptine inhibited the protease activity of ZIKV NS2B-NS3 protein. In addition, bromocriptine exhibited synergistic effect with IFN-α2b against ZIKV replication in cytopathic effect inhibition assay (58).

A number of unique pharmacological characteristics of bromocriptine make it a potentially useful treatment option for ZIKV infection. First, there are different routes of administration of bromocriptine with varying maximum serum concentrations (Cmax) and half-life. Cmax at usual dosage of oral bromocriptine mesylate (5 mg twice a day) is only about 628 ± 375 pg/mL, while the maximum dosage could be increased to 100 mg/day (59). The serum concentration is higher after a single intramuscular injection (50 mg) than that of oral dosing, and the dosages could be increased to up to 100–200 mg (60, 61). The half-life of the slow release process of intramuscular bromocriptine could be as long as 16 days, suggesting that it can be given as a long-acting preparation (60). Second, bromocriptine may be administered per vagina as suppositories or vaginoadhesive discs (62, 63). Vagina administration of bromocriptine is associated with reduced incidence and severity of gastrointestinal and hepatic side effects, as it avoids hepatic first-pass metabolism and is generally well tolerated. Modifications of this vaginal preparation might be especially important for preventing sexual transmission of ZIKV.

9.3.2 Aprotinin

Aprotinin, under the brand name Trasylol, is the small protein bovine pancreatic trypsin inhibitor (BPTI) (64). It is an anti-fibrinolytic molecule, which competitively inhibits several serine proteases. Aprotinin is used as medication administered by injection to reduce bleeding during complex surgery, such as heart and liver surgery. Recently, two independent investigations used this inhibitor to study the crystal structure of ZIKV NS2B-NS3 protein complex (65, 66). In an *in vitro* enzymatic assay, aprotinin potently inhibited ZIKV NS2B-NS3pro enzymatic activity with a K_i of 361 ± 19 nM (65). In the second study, Phoo et al. (2016) designed three different constructs for resolving crystal structure of ZIKV NS2B-NS3. ZIKV NS3 protease (S1-E177) was covalently linked to NS2B cofactor residues (T45-E96) via K126-R130 of NS2B in an eZiPro construct and by G_4SG_4 linker in a gZiPro construct. pETDUET vector with two promoter sites was used for bZiPro, resulting in a unlinked bZiPro construct (Figure 9.4) (66). With an inhibition assay for enzymatic activity, the IC50 values of BPTI were determined to be 350, 76, and 12 nM for eZiPro, gZiPro, and bZiPro, respectively. Based on the IC50 values, BPTI was approximately 30 times less efficient in inhibiting eZiPro than bZiPro, indicating that

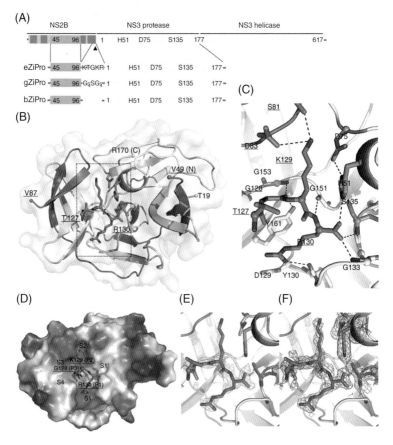

Figure 9.4 Crystal structure of eZiPro in complex with the C terminal TGKR tetrapeptide of NS2B. (A) Full-length NS2B and NS3 proteins and construct designs of eZiPro, gZiPro, and bZiPro. (B) Overall structure of eZiPro showing the TGKR NS2B peptide bound in substrate binding site. NS2B and NS3 are colored in magenta and yellow, respectively. (C) Interactions between viral peptide and residues from protease. (D) Surface charge density view of NS2B-NS3 complex. Residues of substrate binding pockets are labeled. (E) A simulated annealing omit map of the TGKR peptide is contoured at 3σ in green mesh. (F) 2mFo-DFc electron density map contoured at 1σ in blue. *Source:* Extracted from Phoo et al. (2016) (66). (*See insert for color representation of the figure.*)

TGKR peptide competed with BPTI binding to protease pocket by steric hindrance, which was in accordance with the NMR data. BPTI was at least six times less efficiently in inhibiting gZiPro than bZiPro. The higher IC50 value for gZiPro was due to the artificial G_4SG_4 linker between NS2B and NS3, which hindered the substrate from entering the active site (66).

9.3.3   Other Potential Drugs That Target NS2-NS3 Protein Complex

Lei et al. (2016) recently reported a crystal structure of ZIKV NS2B-NS3 protease in complex with a boronate inhibitor, cn-716 (Figure 9.5) (67). The capped dipeptide boronic-acid compound cn-716 was used to obtain the closed conformation of the ZIKV protease. This compound was found to reversibly inhibit ZIKV NS2B-NS3pro with $IC50 = 0.25 \pm 0.02\,\mu M$ and $K_i = 0.040 \pm 0.006\,\mu M$ in the presence of 20% glycerol. In the structure of the complex, the boron atom was covalently linked to the side-chain $O\gamma$ of the catalytic Ser135, and the boronic acid moiety forms a cyclic diester with glycerol, which was continuously present in enzyme preparation, and cryo-protected crystals. In the absence of glycerol, the IC50 value of cn-716 was nearly unchanged ($0.20 \pm 0.02\,\mu M$), but it was noted that through ester formation with larger, more hydrophobic diols or triols, a prodrug could be obtained that could traverse the cellular membrane more readily than free boronic-acid derivatives (67).

Roy et al. (2016) and Lim et al. (2016) identified two small molecules, which targeted ZIKV NS2B-NS3 protease and significantly inhibited its activity (68–70). One is p-nitrophenyl-p-guanidino benzoate (pNGB), an active site inhibitor for both WNV and DENV proteases (71) while the other is quercetin, a natural product rich in many fruits, vegetables, leaves, and grains (Figure 9.6). pNGB showed strong inhibitory activities, with IC50 values of 1.30 ± 0.01 and $0.56 \pm 0.03\,\mu M$, and Ki values of 0.57 ± 0.02 and $0.48 \pm 0.03\,\mu M$, respectively, for linked and unlinked ZIKV NS2B-NS3 complex, which are differentiated by a $(Gly)_4$-Ser-$(Gly)_4$ linker between NS2B and NS3 fragments (68). As for quercetin, it inhibited the enzymatic activity of ZIKV protease with the IC50 and Ki values of $26.0 \pm 0.1\,\mu M$ and $23.0 \pm 1.3\,\mu M$, respectively (70). Different from cn-716, both pNGB and quercetin inhibit ZIKV protease in a noncompetitive mode, suggesting that they are allosteric inhibitors, whose binding pockets have no overlap with the substrate-binding sites (69).

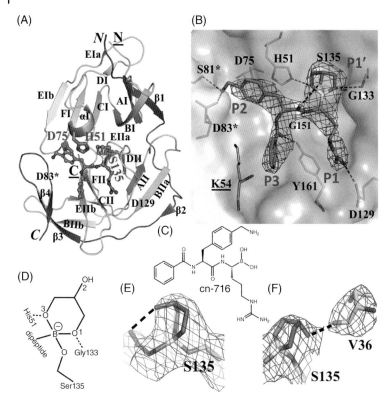

Figure 9.5 Crystal structure of the ZIKV NS2B-NS3pro monomer in complex with cn-716. (A) Ribbon diagram of the complex. NS3pro is shown in light gray and NS2B is shown in dark gray, with secondary-structure elements labeled. The inhibitor cn-716 is shown in sticks with carbon and boron in light gray and white, respectively. (B) Close-up view of the substrate-binding site with cn-716 embedded in. The surfaces of NS2B and NS3pro are shown in white and gray, respectively. (C) Structural formula of cn-716. (D) Schematic drawing and (E) Fobs-Fcalc difference density (2.5σ) for the cyclic diester and its environment in molecule A. (F) Difference density (2.5σ) for the cyclic diester and its environment in molecule B. *Source:* Extracted from Lei et al. (2016) (67).

The virtual screening technique is an economical, reliable, and time-saving method for screening a large set of lead molecules. Using *in silico* screening, Sahoo et al. (2016) screened four FDA-approved drugs (amodiaquine, prochlorperazine, quinacrine, and berberine) against dengue virus infection (72). Among these drugs, berberine showed the highest

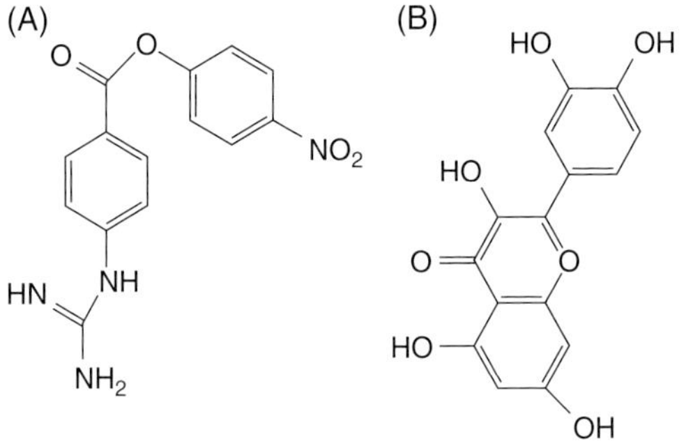

Figure 9.6 Structural formula of two Zika NS2B-NS3pro inhibitors. (A) Chemical structure of p-Nitrophenyl-p-guanidino benzoate. (B) Chemical structure of Qucertin.

binding affinity of −5.8 kcal/mol. It binds around the active site of residues of ZIKV NS3 (Figure 9.7). Furthermore, ZIKV NS3 was used as receptor to screen 4,121 drug-like compounds (small molecules) based on the properties of berberine. More similar compounds were retrieved from ZINC database. Among the best 10 novel drug-like compounds, ZINC53047591(2-(benzylsulfanyl)-3-cyclohexyl-3H-spiro[benzo[h] quinazoline-5,1′-cyclopentan]-4(6H)-one) interacts with NS3 protein with binding energy of −7.1 kcal/mol and formed H-bonds with Ser135 and Asn152 amino acid residues (Figure 9.8) (72).

Although these *in silico/in vitro* studies extend platform for developing anti-viral competitive inhibitors against ZIKV infection, further investigations using *in vitro* cell-based assay and *in vivo* animal experiments are required to confirm the antiviral functionality of these identified drug candidates.

9.4 Drugs That Target ZIKV NS5 RNA-Dependent RNA Polymerase Activity

ZIKV NS5 consists of two functional domains: the N-terminal domain contains the methyltransferase and guanylyltransferase activities that contribute to the 5′ cap structure formation, while the C-terminal domain contains the RNA-dependent RNA polymerase (RdRp) activity that is responsible for viral genome synthesis (73–79). Because human cells do not have RdRp activity, NS5 RdRp domain is an

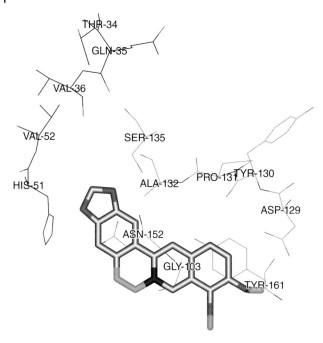

Figure 9.7 Molecular docking complex of berberine with nonstructural 3 (NS3) protein of Zika virus. *Source:* Extracted from Sahoo et al. (2016) (73). (*See insert for color representation of the figure.*)

attractive target for designing antiviral drugs that might cause fewer deleterious side effects on patients (80, 81).

9.4.1 Sofosbuvir

Using small molecule compounds to block NS5 RdRp domain and its enzymatic function has been demonstrated to be a successful antiviral strategy against other related RNA viruses, including HCV. Sofosbuvir, sold under the brand name Sovaldi, is an RdRp inhibitor approved by FDA for the treatment of HCV infection (82). Sofosbuvir is a uridine nucleotide analog prodrug, which is metabolized to 2-F-2-C-methyluridine monophosphate and converted to the active triphosphate form within hepatocytes to target viral RNA polymerase by acting as a chain terminator (83–85).

Recently, anti-ZIKV activity of sofosbuvir has been reported (86–88). The residues critical for RdRp activity are conserved among

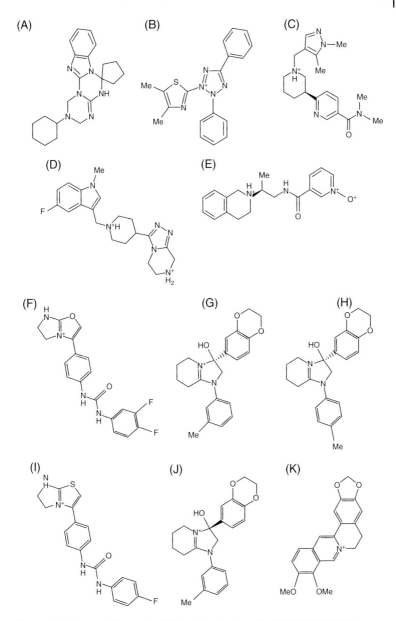

Figure 9.8 Structural formula of 10 best lead molecules. (A) ZINC53047591, (B) ZINC13510840, (C) ZINC19705600, (D) ZINC19711173, (E) ZINC25634061, (F) ZINC98342354, (G) ZINC02974658, (H) ZINC04086851, (I) ZINC98342344, (J) ZINC02974656, and (K) berberine (ZINC03779067). *Source:* Extracted from Sahoo et al. (2016) (72).

different viral species and strains, including African ZIKV strain MR766 and those currently circulating, DENV and HCV (86). Cell culture experiments showed that sofosbuvir inhibited ZIKV replication in human hepatoma (Huh-7) and placenta (Jar) cell lines, and fetal-derived hindbrain and cerebral cortex neuronal stem cells (87), as well as neocortical and spinal cord neuroepithelial stem cells (88). In addition, brain organoids, a three-dimensional model for examining the impact of ZIKV on early brain development, were used to assess sofosbuvir's antiviral activity. A pronounced reduction of ZIKV production was observed following sofosbuvir treatment (86). *In vivo* study in mice showed that oral administration of sofosbuvir with a physiologically relevant dose ($\sim$33 mg/kg/day $\times$ 7 days) prevented weight loss and death in 50.0% of treated mice (87). Interestingly, it seemed sofosbuvir was more effective in reducing ZIKV infectivity than viral RNA levels in the supernatant. In other words, the genomic RNA-containing virions might be produced in the presence of sofosbuvir treatment, but these particles were unable to efficiently infect new cells and cause cytopathogenicity (86).

Although sofosbuvir is an FDA-approved medication for treating chronical HCV infection, the immediate use of sofosbuvir as monotherapy for preventing microcephaly and other congenital malformation seems unlikely. Sofosbuvir is a pregnancy category B FDA-approved drug. Although animal studies with sofosbuvir have not shown any evidence of fetal harm (89), there is no controlled result in human pregnancy. Another key factor that might affect the use of sofosbuvir for therapeutically treating ZIKV infection is the price. In the United States, a course of treatment with Sovaldi (sofosbuvir) costs approximately $84,000. This high cost of sofosbuvir might put it out of reach for many low-income individuals.

9.4.2 Other Nucleoside Inhibitors of ZIKV

BCX4430, an adenosine nucleoside analog, functions as a selective antiviral inhibitor targeting viral RdRp, and is currently in clinical stage development (90). It has broad-spectrum activity against a wide range of RNA viruses, including YFV, Marburg virus, and Ebola virus (90, 91). Julander et al. (2017) tested BCX4430 against African (MR766), Asian (P 6-740), and American lineage of ZIKV (PRVABC59) in cytopathic effect inhibition and virus yield reduction assays in various cell lines (92). BCX4430 consistently reduced viral cytopathic effect (CPE) induced by ZIKV in RD, Huh-7, and Vero76 cell lines with EC50 values in the low µg/mL range (from 3.8 to 11.7 µg/mL) and favorable selective index (SI) values (from 5.5 to 20.5). Furthermore, Julander et al. (2017) used A129 mouse model to evaluate its efficacy *in vivo* (92) and found that BCX4430 treatment significantly improved the survival of ZIKV-infected animals, protecting them against virus-led weight loss and reducing the viral RNA level in the serum, indicating the potential use of BCX4430 in treatment of ZIKV infection (92). In this study, the anti-ZIKV activity of Ribavirin, which was previously showed to have broad-spectrum activity in this cell line against YFV and WNV (93, 94), was also evaluated both *in vitro* and *in vivo*. Although human cell lines (RD and Huh-7) confirmed its anti-ZIKV activity, Ribavirin failed to improve the outcome of AG129 mice infected by ZIKV infection (92).

NITD008, an adenosine nucleoside analog inhibitor, contains a carbon substitution of N-7 of the purine and an acetylene at the 2′ position of ribose (95). Its triphosphate form can compete with natural adenosine triphosphate substrates for incorporation into the growing RNA chain leading to termination of RNA elongation. NITD008 was reported to exhibit potent antiviral activity against DENV *in vitro* and *in vivo* (95). It also inhibited the replication of several other flaviviruses and enteroviruses (96–98). NITD008 effectively inhibited the progeny viral titers of ZIKV strains GZ01/2016 (88) and FSS13025/2010 (99) with EC50 values calculated to be 241 nM and 137 nM, respectively (100). The *in vivo* therapeutic effect of NITD008 was assessed using the A129 mouse model, which lacked type I IFN signaling (37). Treatment with 50 mg/kg NITD008 protected 50.0% of the infected mice from death, comparing a 100.0% mortality rate of mock treated group. All the infected mice of control group developed neurological symptoms, including hind limb weakness and paralysis while the survival mice from NITD008 treatment group did not develop any neurological symptoms (100).

7DMA, 7-deaza-2′-C-methyladenosine, is an adenosine analog. It was originally developed by Merck Research Laboratories as an inhibitor for hepatitis C virus infection (101). Despite promising results in animal studies (102, 103), it was ultimately unsuccessful in clinical trials (104). Subsequently, it was widely used in antiviral research, and was shown to inhibit the replication of a range of viruses, including DENV, tick-borne encephalitis virus, and poliovirus (105–108). 7DMA significantly inhibited the replication of ZIKV MR766 strain with EC50 values ranging between 5 μM and 20 μM as assessed by both virus yield reduction assay and CPE-reduction assay (109). In the *in vivo* study, infected AG129 mice were treated once daily with 50 mg/ kg/day of 7DMA or vehicle via oral gavage. In two independent studies, mice treated with 7DMA had markedly delayed virus-induced disease progression with a mean day of euthanasia (MDE) of 23.0 and 24.0 compared to those treated with vehicle (MDE of 14.0 and 16.0 days, respectively). Besides 7DMA, the potential anti-ZIKV activities of several other known (+)ssRNA virus inhibitors: 2'-C-methylcytidine, 7-deaza-2'-C-methyladenosine, ribavirin, T-705 and its analogue T-1105 were also evaluated (109). All of these compounds resulted in a selective, dose-dependent inhibitory effect on ZIKV replication.

In addition, using well-designed *in vitro* assays, several other inhibitors of nucleoside analogs have been identified as promising therapeutic candidates for further development of specific antiviral agents against ZIKV (110, 111). Eyer et al. (2016) tested a series of 29 nucleoside analogues at a concentration of 50 μM for their abilities to inhibit ZIKV-induced CPE in Vero cells (110). Five of the nucleoside analogues, 7-deaza-2'-C-methyladenosine (7-deaza-2'-CMA), 2'-C-methyladenosine (2'-CMA), 2'-C-methylcytidine (2'-CMC), 2'-C-methylguanosine (2'-CMG), and 2'-C-methyluridine (2'-CMU), were found to inhibit ZIKV-induced CPE in cell culture and significantly reduce cell death compared with mock-treated ZIKV-infected cells (Table 9.1). Further evaluation showed that all 2'-C-methylated derivatives reduced the viral titers in a dose-dependent manner with mean EC50 values ranging from 5.26 to 45.45 μM. Their antiviral activities were substantially affected by the identities of the hetero-cyclic base moieties (110).

In another study, six ATP analogs were tested for inhibition of ZIKV RdRp activities in an *in vitro* enzymatic activity assay utilizing purified recombinant RdRp from ZIKV MR766 strain (Figure 9.9) (111). ATP analogs 1, 2, 3, and 6 exhibited significant inhibition of ZIKV RdRp with IC50 values calculated to be 5.6 μM, 7.9 μM, 6.4 μM, and 2.7 μM, respectively. In contrast, compound 4 and 5 were virtually inactive. Compound 1 and 2 are 2'-C-methylated nucleotides. Eyer et al. (2016) showed that the parent nucleosides of these two compounds exerted significant inhibitory effects on ZIKV replication in cell-based assay (110). Both compound 3 and 6 are conformationally locked nucleotides with 3 in the Northern-like conformation and 6 in the Southern-like

Table 9.1 ZIKV Inhibition and Cytotoxicity Characteristics of Selected Nucleoside Analogues.

Compound	Structure	EC50, μM, Mean ± SD[a,b]	Apparent EC50, μM, Mean ± SD[a,c]	CC50, μM[a]	SI[d]
2′-C-methyladenosine		5.26 ± 0.12	26.41 ± 1.14	>100[f]	>19.01
7-deaza-2′-C-methyladenosine		8.92 ± 3.32	42.87 ± 1.88	>100	>11.21
2′-C-methylguanosine		22.25 ± 0.03	71.23 ± 4.11	>100	>4.49
2′-C-methylcytidine		10.51 ± 0.02	61.93 ± 3.76	>100[g]	>9.51

(Continued)

Table 9.1 (Continued)

Compound	Structure	EC50, μM, Mean ± SD[a,b]	Apparent EC50, μM, Mean ± SD[a,c]	CC50, μM[a]	SI[d]
2′-C-methyluridine		45.45 ± 0.64	>100[e]	>100	>2.20

Source: Extracted from Eyer et al. (2016) (110).

Notes

Abbreviations: CC50, 50% cytotoxic concentration; EC50, 50% effective concentration; SD, standard deviation; ZIKV, Zika virus.

[a] Determined from three independent experiments.

[b] Calculated as a 50% reduction of virus titers, using the Reed-Muench method.

[c] Calculated as a point of inflection from dose-response curves, using log-transformed viral titers.

[d] The selectivity index (SI) is calculated as CC50/EC50.

[e] Treatment with 100 μM of 2′-CMU reduced virus titers 10^2-fold compared to a mock-treated culture.

[f] Treatment of cell culture with 2′-CMA at a concentration of 100 μM led to a reduction in cell viability to 79.1%.

[g] Treatment of cell culture with 2′-CMC at concentration of 100 μM led to a reduction in cell viability to 69.3%.

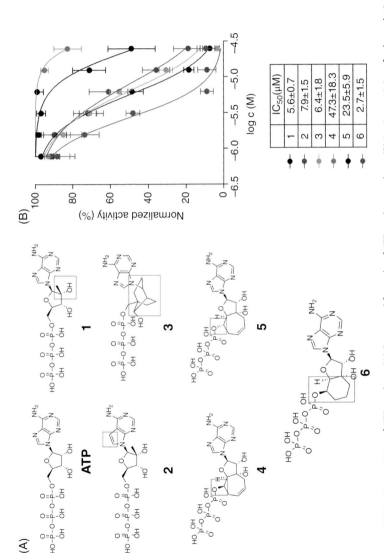

Figure 9.9 *In vitro activity of ATP analogs. (A) Structural formula of ATP and its analogs. (B) In vitro activity of selected triphosphates. Source: Extracted from Hercik et al. (2017) (111). (See insert for color representation of the figure.)*

conformation of the sugar moiety. Both North- and South-locked nucleotides act as potent inhibitors of ZIKV RdRp though the mode of action is still elusive (111).

9.5 Neutralizing Antibodies That Target ZIKV Structural Protein

Mature flavivirus particles contain 180 copies of the envelope (E) protein and membrane (M) protein on the envelope, and display an icosahedral arrangement in which 90 E protein dimers completely cover the viral surface (112–114). Flavivirus E protein is responsible for virus entry and represents a major target for neutralizing antibodies. A number of potent flavivirus type-specific neutralizing antibodies specific to E protein domain III have been identified while cross-neutralizing antibodies target the fusion loop region or E-dimer region are less potent (115–119). Moreover, several previous reports have shown the *in vivo* protective efficacies of antibodies to fusion loop epitope (FLE) (115, 120–123).

Dai et al. (2016) reported the structures of ZIKV E protein at 2 Å and in complex with a flavivirus broadly protective murine antibody 2A10G6 at 3 Å (124). This antibody exhibited *in vitro* neutralizing activity against DENV 1–4 and WNV (120). 2A10G6 binds to the ZIKV E protein with a high binding affinity of 2.7 nM, and neutralizes the infection of a currently circulating ZIKV Asian strain SZ01 (125). The antibody exhibited neutralizing activity with a 50% plaque reduction neutralizing titer ($PRNT_{50}$) of 249 µg/mL. A single dose of 500 µg of 2A10G6 antibody provided complete protection *in vivo* against ZIKV infection in the A129 mouse model in a 60.0% mortality setting of the control group (124). Furthermore, the structure of ZIKV E/2A10G6 complex showed that the antibody bound to the tip of E protein domain II at a perpendicular angle, embedding the fusion loop of E protein in a hydrophobic hole formed by both the heavy chain and light chain. The hydrophobic residues Trp101, Leu107, and Phe108 form the majority of interactions between 2A10G6 and ZIKV E protein, and are highly conserved in the majority of flaviviruses, providing an explanation for the broadly protective capacity of 2A10G6 (Figure 9.10) (124). Since these three residues are highly conserved in all of the isolated ZIKV strains, antibody 2A10G6, especially after being humanized, might represent an auspicious therapeutic agent for the treatment of the ZIKV infection.

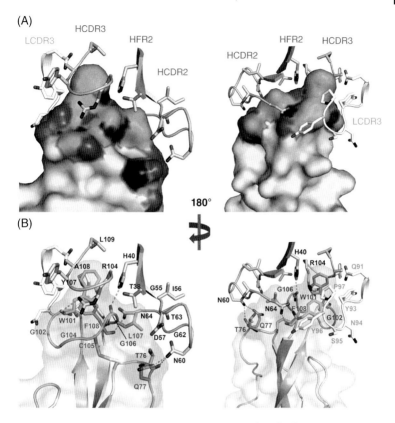

Figure 9.10 Interactions between ZIKV E protein and antibody 2A10G6. (A) Four polypeptide elements of antibody 2A10G6 surround the domain II tip of ZIKV E protein. Three polypeptide elements from the heavy chain are shown in light gray and one from the light chain is shown in white. (B) The heavy chain of antibody 2A10G6 forms five hydrogen bonds (dashed lines) with the ZIKV E protein. *Source:* Extracted from Dai et al. (2016) (124).

9.6 Drugs That Inhibit ZIKV Infection by Targeting Host Machinery

In contrast to targeting viral proteins, targeting the host factors has some advantages since the host targets are less susceptible to mutations. Therefore, discovery of antiviral drugs that target host factors may result in a better treatment and control of ZIKV infections.

9.6.1 Bithionol

Bithionol was previously approved to treat helminthes, and then replaced by the more effective anti-helminthic, Praziquantel (126). Although the exact mechanism of action is unclear, it is believed that the drug inhibits oxidative phosphorylation of the parasites. Bithionol was also used as a bacteriostatic additive in soaps and cosmetic products until the FDA banned it in 1967 for its photosensitizing effects. Currently, Bithionol is still used for treating helminthes in some Asian and European countries, and for mouth and throat disorders in Argentina (127).

Leonardi et al. (2016) uncovered five Bithionol targets: caspase-1, -3, -6, -7, and -9, which make Bithionol a poly-pharmacological drug (128). In this study, it was found that Bithionol is capable of reducing the pathogenicity of a wide range of pathogenic agents, including bacterial toxins, ricin, and ZIKV. ZIKV infection causes cell death by inducing host caspase-3 and neuronal apoptosis during its propagation (129, 130). In addition, caspases facilitate virus replication and propagation by cleaving various viral proteins, affecting viral protein localization, and promoting viral genome replication and virus assembly (131, 132).

In a study, two ZIKV strains, Senegal strain DAKAR D41525 and Puerto Rico strain PRVABC59, were selected to test Bithionol's antiviral activity. With an *in vitro* immuno-staining assay, Bithionol was found to inhibit the viral yield of Puerto Rico strain in Vero E6 cells with an EC50 of 6.7 µM, as well as Senegal strain in Vero E6 and human astrocytes with EC50 values of 5.5 µM and 6.3 µM, respectively (128).

Bithionol, as a previously approved drug, has a well-established safety and pharmacokinetic profiles in patients, and could be rapidly made available for a new indication. Moreover, after oral administration of 50 mg/kg of Bithionol in humans, serum concentrations of the drug could reach the range from 225 to 480 µM, which is approximately 50-fold higher than the EC50 values of Bithionol for ZIKV infection determined by the cell-based tests (128), suggesting that it is possible to achieve effective anti-ZIKV Bithionol doses in blood by oral administration. Besides, Bithionol is able to cross the blood-brain barrier and placenta barrier. It has been used to treat paragonimiasis in the population of children and pregnant women (133). All of these features of Bithoinol make it a promising anti-ZIKV drug candidate.

9.6.2   Gemcitabine

Gemcitabine, under the brand name Gemzar, a chemotherapy medication, is used to treat numerous types of cancer, including ovarian cancer, non–small-cell lung cancer, and pancreatic cancer. It interferes with *de novo* synthesis of pyrimidine and inhibits replication of FLUAV, FLUBV, SINV, and HSV-1 (38, 41). In the aforementioned study of Kuivanen et al. (2017), gemcitabine rescued infected human RPE cells from cell death induced by ZIKV (FB-GWUH-2016 strain) (45). Similar to obatoclax and SaliPhe, gemcitabine exhibited dose-dependent anti-ZIKV activity, inhibiting viral protein expression, viral genomic RNA synthesis, and progeny virion production. All of three compounds prevented caspase-3, -8, and -7 activation in ZIKV infected cells. However, gemcitabine is different from obatoclax and SaliPhe in two key aspects:

1) Gemcitabine can protect cells from ZIKV-induced cell death even when given at 10 h.p.i., indicating that it does not affect ZIKV entry and trafficking.
2) Treatment of obatoclax and SaliPhe but not gemcitabine prevented ZIKV-induced antiviral responses, including the activation of innate immunity system, IFN signaling, and RIG-I and MDA5 pathways (45).

Taken together, gemcitabine might inhibit ZIKV infection by interfering with transcription of viral RNA.

9.6.3   Inhibitors of Pyrimidine Synthesis: 6-Azauridine, CID91632869 and Brequinar

6-azauridine, a uridine analog, inhibits virus infection by interfering with pyrimidine biosynthesis, thereby preventing formation of cellular nucleic acids (134, 135). In the study of Adcock et al. (2017), its EC50 values were determined to be 3.91 and 3.18 μM in a CPE-based assay for PRVABC59 and MR766, respectively, indicating effective anti-ZIKV activities (136). However, with the same assay, the antiviral activities of 6-azauridine were examined against two alphaviruses (VEEV TC-83 and CHIKV) and two other flaviviruses (WNV and JEV). Interestingly, 6-azauridine had no antiviral effect on both alphaviruses, while all flaviviruses showed a high susceptibility to 6-azauridine treatment (136).

CID91632869 is an inhibitor of pyrimidine synthesis and a broad-spectrum antiviral compound that was initially discovered from a structure-activity relationship study of a hit compound from a high-throughput screen (HTS) of Venezuelan equine encephalitis virus (VEEV) (137).

Brequinar is a synthetic quinolinecarboxylic acid analogue with antineoplastic properties. Brequinar inhibits dihydroorotate dehydrogenase, thereby blocking *de novo* pyrimidine biosynthesis (138). In a recent HTS study, Adcock et al. (2017) evaluated the anti-ZIKV activities of these two pyrimidine inhibitors and found that both compounds showed prominent antiviral activity against all flaviviruses tested (ZIKV, WNV, and JEV) without any cytotoxicity (136). As for their anti-ZIKV efficiencies, brequinar showed submicromolar EC50, and CID 91632869 showed EC50 values of 1.09 and 2.17 μM for ZIKV MR766 and PRVABC59 strains, respectively. The mechanism of action of CID 91632869 or brequinar could be two-fold: (1) inhibition of pyrimidine synthesis; and (2) induction of antiviral gene products. Based on the current understanding of mechanism for these inhibitors of pyrimidine synthesis, the major antiviral effects of these compounds come from induction of antiviral gene expression (139). It seems that flaviviruses, including ZIKV, might not have mechanisms for evading innate antiviral gene products induced by inhibitors of pyrimidine synthesis.

Although inhibitors of pyrimidine synthesis show good antiviral activities *in vitro*, *in vivo* test using cotton rats or mice models have failed to show antiviral activity (140–143). High level of uridine in the blood might overcome the compound effect by supplying pyrimidines via a salvage pathway (143). In addition, species specificity of activating innate immunity genes by pyrimidine depletion could be another reason for the failure (137). Developing a proper *in vivo* animal model for evaluating inhibitors of pyrimidine synthesis is key for their therapeutic translation into potential antiviral agents.

9.6.4 Cyclin-Dependent Kinase Inhibitor (CDKi): PHA-690509

PHA-690509 is an investigational compound that functions as an ATP competitive CDKi. Xu et al. (2016) demonstrated that PHA-690509 inhibited three strains of ZIKV (MR766, FSS13025, and PRVABC59), determined by NS1 expression in glioblastoma SNB-19 cell lines, in a dose-dependent fashion (52). PHA-690509 inhibited viral RNA expression with an IC50 value of 1.72 μM, and suppressed the production of infectious ZIKV particles at sub-micromolar concentrations. In addition, both time-of-addition and time-course experiments suggested PHA-690509 inhibited ZIKV infection at a post-entry stage (52). Further, the effectiveness of PHA-690509 was examined in hNPCs and astrocytes, both of which were target cells for ZIKV infection in the fetal brain (44). PHA-690509 treatment inhibited ZIKV production with an IC50 value of ~0.2 μM, and showed only minimal cell toxicity at a level <0.3 μM. As for hNPCs, ZIKV infection led to a drastic reduction in hNPC proliferation, which was partially rescued by PHA-690509; and treatment with PHA-690509 at 1 μM alone had minimal effect on hNPC proliferation in brain organoids (52).

Pharmacological CDKis inhibit replication of diverse viruses *in vitro*, including HIV and HSV-1; and depletion of CDK9 impairs influenza A virus replication (144–150). Due to the intimate connection between ZIKV infection and cell cycle regulation (129), Xu et al. (2016) further tested 27 additional structurally distinct CDKis for inhibition of ZIKV infection, and identified 9 effective compounds using the SNB-19 cell model (52). Analyses of ZIKV production determined the IC50 values to be at sub-micromolar concentrations for 8 of 9 CDKis tested. Amongst these CDKis, seliciclib and RGB-286147 were the most effective ones, with IC50 values of 24 nM and 27 nM, respectively.

Since 10 structurally unrelated CDKis have shown inhibitory effect on ZIKV infection and replication, CDKs may play essential roles in the replication cycle of ZIKV in humans. Further investigations are needed to identify the specific cellular CDKs in the host that are important for ZIKV infection, and to reveal the mechanism of action of these inhibitors. Although many CDKis have been evaluated in clinical trials for various cancers, cystic fibrosis, and Cushings disease, these compounds might not be suitable for pregnant women due to their potential hazardous effects on the fetus. Future *in vivo* animal studies, including both rodents and primates, seem critical to test the efficacies and toxicities of the identified compounds (52).

9.7 Drugs That Show Neuroprotective Activity but Do Not Suppress ZIKV Replication: Emricasan

Emricasan, also named IDN-6556 or PF-03491390, is an inhibitor of activated caspases with sub- to nanomolar activity *in vitro* (151, 152). Emricasan was developed for the treatment of liver disease, and is currently evaluated in phase II clinical trials for the reducing hepatic injury and liver fibrosis caused by chronic HCV infection (153, 154). In the latest screening for compounds that reduced virally induced caspase activation and apoptosis by directly preventing ZIKV-induced cell death, Xu et al. (2016) identified emricasan as the most potent anti-cell-death compound (52). Emricasan effectively protected glioblastoma SNB-19 cells from cell death induced by three ZIKV strains: MR766, FSS13025, and PRVABC59 with IC50 values of 0.13 to 0.9 µM in both caspase activity and cell viability assays. Emricasan was also effective in two other tested cell lines: astrocytes and human neural progenitor cells (hNPCs). Although emricasan can protect neural cells from ZIKV-induced death, it does not suppress ZIKV replication (52).

In a two-drug combination experiment using emricasan and the aforementioned CDKi PHA-690509, Xu et al. (2016) found that the combination treatment led to an additive effect in inhibiting caspase-3 activity and in preserving astrocyte viability after ZIKV infection. Importantly, emricasan did not interfere with the anti-ZIKV activity of CDKi in the combination treatment (52). In a sequential treatment experiment, ZIKV-infected hNPCs were treated with emricasan for 72 hours followed by niclosamide treatment for 48 hours. Such combination of treatment resulted in the recovery of ZIKV-negative hNPCs, indicating a beneficial effect of "buying time" by inhibiting apoptosis to allow the infected cells to recover (52).

As an investigational drug, emricasan was well tolerated in human clinical trials, and there was no significant adverse event (154). A previous report showed that the overall and maximum concentrations of emricasan (orally, twice/day, ×4 days) in human blood were 1.90 µg/mL (3.35 µM) and 2.36 µg/mL (4.15 µM), respectively, which were about 10-fold higher than the IC50 for inhibiting the increased caspase-3 activity and cell death caused by ZIKV infection *in vitro* (52). It will be critical to evaluate the efficacy of emricasan in *in vivo* studies in the future.

9.8 Other Drugs That Inhibit ZIKV Infection Identified from a Screening of FDA-Approved Drugs

Infection of flaviviruses requires a large number of human host factors, and many of these factors are the targets of approved drugs. An expeditious way to identify anti-ZIKV therapeutic candidates is to repurpose previous approved drugs (31, 52). Barrow et al. (2016) screened a library of 774 FDA-approved therapeutic drugs and identified 24 drugs with validated anti-ZIKV activities (Table 9.1) (31).

Among the identified potent drugs are ivermectin, mycophenolic acid (MPA), and daptomycin. MPA and daptomycin showed EC50 values of between 0.1 and 1.0 μM, respectively, while the EC50 value for ivermectin was determined to be between 1 and 10 μM (31). Other notable drugs that reduced ZIKV infection without strongly reducing cell numbers were sertraline-HCl, pyrimethamine, and cyclosporine A. Further validations of these drugs were conducted using HeLa cells, hNSC cell line (K048), JEG3 cells derived form a placental choriocarcinoma, and primary human amnion epithelial cells (hAECs) derived from the lining of the amniotic sac (31). All these cells were effectively infected by ZIKV strain MEX_I_7. In HeLa cells, all drugs tested reduced ZIKV infection at 10 μM concentrations. In JEG3 cells, ivermectin and MPA showed potent antiviral effects, with MPA being the most potent one, reducing ZIKV infection at 1 and 10 μM. Using an hNSC cell line, at concentration of 1 μM, only MPA reduced virus infection. At 10 μM concentrations, MPA, cyclosporine A, or ivermectin significantly blocked ZIKV MEX_I_7 infection; however, ivermectin, and to a less extent, cyclosporine A reduced the cell number and caused cell toxicity. Lastly, when the anti-ZIKV activities of these drugs were tested in hAECs, none of these drugs caused substantial reduction of cell numbers at 16 μM concentrations. In contrast, moderate inhibition of ZIKV infectivity by daptomycin, strong inhibition by ivermectin and sertraline at 16 μM, and strong inhibition by MPA at 1.6 μM were observed (31).

Ivermectin, under brand names Heartgard, Sklice, and Stromectol, is a medication that is effective against many types of parasites (155). It has previously been shown to inhibit VEEV, CHIKV, and several flaviviruses (156–158). MPA, sold under brand names CellCept and Myfortic, is an immunosuppressant drug, used to prevent rejection in organ transplantation by inhibiting an enzyme needed for the growth of T cells and B cells (159, 160). MPA has been shown to inhibit replication of DENV (161–164). Daptomycin, under brand name Cubicin, is a lipopeptide antibiotic that is used to treat systemic and life-threatening infections caused by Gram-positive organisms (165). Daptomycin has not previously been associated with any antiviral activity. Daptomycin is able to insert into cell membranes rich in phosphatidylglycerol (PG) (166), suggesting an effect on PG-rich late endosomal membranes, which are critical for flavivirus entry (167).

Table 9.2 Candidate Anti-ZIKV Drugs and Considerations for Use in Pregnancy.

Drug Name	Pregnancy Category	Other Considerations and Notes
Auranofin	C	Inform women of childbearing potential of the potential risk of therapy during pregnancy.
Azathioprine	D	Contraindicated for use in pregnant women with rheumatoid arthritis.
Bortezomib	D	Women of reproductive potential should avoid becoming pregnant while on therapy.
Clofazimine	C	Some animal studies have failed to reveal evidence of teratogenicity, but studies done at high doses have demonstrated fetotoxicity. There are no controlled data in human pregnancy.
Cyclosporine A	C	Advise of the potential risks if used during pregnancy.
Dactinomycin	D	
Daptomycin	B	
Deferasirox	C	
Digoxin	C	Concentrations with anti-ZIKV activity may be toxic.
Fingolimod	C	A pregnancy registry has been established to collect information about the effect of this drug during pregnancy.
Gemcitabine·HCl	D	
Ivermectin	C	
Mebendazole	C	Inform of potential risk to fetus if taken during pregnancy, especially during first trimester.
Mefloquine·HCl	B	
Mercaptopurine Hydrate	D	
Methoxsalen	C	Usually given in combination with UV radiation therapy.
Micafungin	C	

Table 9.2 (Continued)

Drug Name	Pregnancy Category	Other Considerations and Notes
Mycophenolate Mofetil	D	Boxed warning: Use during pregnancy is associated w/increased risks of 1st trimester pregnancy loss and congenital malformations; counsel females of reproductive potential regarding pregnancy prevention and planning.
Mycophenolic Acid	D	Boxed warning: Use during pregnancy is associated w/increased risks of pregnancy loss and congenital malformations; counsel females of reproductive potential regarding pregnancy prevention and planning.
Palonosetron·HCl	B	Drug interaction with SSRIs (Sertraline) causing serotonin syndrome.
Pyrimethamine	C	Women of reproductive potential should avoid becoming pregnant while on therapy.
Sertraline·HCl	C	Boxed warning: Antidepressants increased the risk of suicidal thinking and behavior (suicidality) in children, adolescents, and young adults in short-term studies of major depressive disorder (MDD) and other psychiatric disorders. Consider tapering dose during third trimester of pregnancy.
Sorafenib Tosylate	D	Inform that the drug may cause birth defects or fetal loss during pregnancy; instruct both males and females to use effective birth control during treatment and for at least 2 weeks after stopping therapy. Instruct to notify physician if patient becomes pregnant while on therapy.

Source: Extracted from Barrows et al. (2016) (31).

Sertraline (under brand name Zoloft), a serotonin reuptake inhibitor, is primarily prescribed for major depressive disorder, obsessive-compulsive disorder, panic disorder, and social anxiety disorder (168). Pyrimethamine, sold under the brand name Daraprim, is a medication used with leucovorin to treat toxoplasmosis and cystoisosporiasis. Cyclosporine A has been approved for treating graft-versus-host disease in bone-marrow transplantation and preventing rejection of kidney, heart, and liver transplants through lowering the activities of T cells and their immune responses (169). Cyclosporine A has previously been documented to inhibit flavivirus infection (170).

Besides the antiviral activities of these drugs, three other factors might affect their potential therapeutic applications for treating ZIKV infection:

1) Given their safety profiles, many of these drugs have been used during pregnancy for other indications, both in the United States as well as globally (Table 9.2).
2) Many of the drugs shown to have anti-ZIKV activities could have untoward effect, particularly in the context of pregnancy.
3) The pharmacokinetics, such as the maximum plasma concentrations (C_{max}) and the duration to achieve C_{max} (t_{max}) have been reported for some of these drugs (31).

References

1 Calland N, Albecka A, Belouzard S, Wychowski C, Duverlie G, Descamps V, Hober D, Dubuisson J, Rouille Y, Seron K. 2012. (-)-Epigallocatechin-3-gallate is a new inhibitor of hepatitis C virus entry. Hepatology **55**:720–729.

2 Isaacs CE, Wen GY, Xu W, Jia JH, Rohan L, Corbo C, Di Maggio V, Jenkins EC, Jr., Hillier S. 2008. Epigallocatechin gallate inactivates clinical isolates of herpes simplex virus. Antimicrob Agents Chemother **52**:962–970.

3 Nance CL, Siwak EB, Shearer WT. 2009. Preclinical development of the green tea catechin, epigallocatechin gallate, as an HIV-1 therapy. J Allergy Clin Immunol **123**:459–465.

4 Song JM, Lee KH, Seong BL. 2005. Antiviral effect of catechins in green tea on influenza virus. Antiviral Res **68**:66–74.

5 Huang HC, Tao MH, Hung TM, Chen JC, Lin ZJ, Huang C. 2014. (-)-Epigallocatechin-3-gallate inhibits entry of hepatitis B virus into hepatocytes. Antiviral Res **111**:100–111.

6 Kim M, Kim SY, Lee HW, Shin JS, Kim P, Jung YS, Jeong HS, Hyun JK, Lee CK. 2013. Inhibition of influenza virus internalization by (-)-epigallocatechin-3-gallate. Antiviral Res **100**:460–472.

7 Carneiro BM, Batista MN, Braga AC, Nogueira ML, Rahal P. 2016. The green tea molecule EGCG inhibits Zika virus entry. Virology **496**:215–218.

8 Yamaguchi K, Honda M, Ikigai H, Hara Y, Shimamura T. 2002. Inhibitory effects of (-)-epigallocatechin gallate on the life cycle of human immunodeficiency virus type 1 (HIV-1). Antiviral Res **53**:19–34.

9 Browning DJ. 2014. Hydroxychloroquine and Chloroquine Retinopathy. Springer, New York, NY, USA.

10 Tsai WP, Nara PL, Kung HF, Oroszlan S. 1990. Inhibition of human immunodeficiency virus infectivity by chloroquine. AIDS Res Hum Retroviruses **6**:481–489.

11 Ooi EE, Chew JS, Loh JP, Chua RC. 2006. In vitro inhibition of human influenza A virus replication by chloroquine. Virol J **3**:39.

12 Farias KJ, Machado PR, da Fonseca BA. 2013. Chloroquine inhibits dengue virus type 2 replication in Vero cells but not in C6/36 cells. ScientificWorldJournal **2013**:282734.

13 Zhu YZ, Xu QQ, Wu DG, Ren H, Zhao P, Lao WG, Wang Y, Tao QY, Qian XJ, Wei YH, Cao MM, Qi ZT. 2012. Japanese encephalitis virus enters rat neuroblastoma cells via a pH-dependent, dynamin and caveola-mediated endocytosis pathway. J Virol **86**:13407–13422.

14 Boonyasuppayakorn S, Reichert ED, Manzano M, Nagarajan K, Padmanabhan R. 2014. Amodiaquine, an antimalarial drug, inhibits dengue virus type 2 replication and infectivity. Antiviral Res **106**:125–134.

15 Delvecchio R, Higa LM, Pezzuto P, Valadao AL, Garcez PP, Monteiro FL, Loiola EC, Dias AA, Silva FJ, Aliota MT, Caine EA, Osorio JE, Bellio M, O'Connor DH, Rehen S, de Aguiar RS, Savarino A, Campanati L, Tanuri A. 2016. Chloroquine, an Endocytosis Blocking Agent, Inhibits Zika Virus Infection in Different Cell Models. Viruses **8**.

16 Wolfe MS, Cordero JF. 1985. Safety of chloroquine in chemosuppression of malaria during pregnancy. Br Med J (Clin Res Ed) **290**:1466–1467.

17 Law I, Ilett KF, Hackett LP, Page-Sharp M, Baiwog F, Gomorrai S, Mueller I, Karunajeewa HA, Davis TM. 2008. Transfer of chloroquine and desethylchloroquine across the placenta and into milk in Melanesian mothers. Br J Clin Pharmacol **65**:674–679.

18 Barbosa-Lima G, da Silveira Pinto LS, Kaiser CR, Wardell JL, De Freitas CS, Vieira YR, Marttorelli A, Cerbino Neto J, Bozza PT, Wardell SM, de Souza MV, Souza TM. 2017. N-(2-(arylmethylimino) ethyl)-7-chloroquinolin-4-amine derivatives, synthesized by thermal and ultrasonic means, are endowed with anti-Zika virus activity. Eur J Med Chem **127**:434–441.

19 Pukrittayakamee S, Imwong M, Looareesuwan S, White NJ. 2004. Therapeutic responses to antimalarial and antibacterial drugs in vivax malaria. Acta Trop **89**:351–356.

20 Brickelmaier M, Lugovskoy A, Kartikeyan R, Reviriego-Mendoza MM, Allaire N, Simon K, Frisque RJ, Gorelik L. 2009. Identification and characterization of mefloquine efficacy against JC virus in vitro. Antimicrob Agents Chemother **53**:1840–1849.

21 Yan KH, Lin YW, Hsiao CH, Wen YC, Lin KH, Liu CC, Hsieh MC, Yao CJ, Yan MD, Lai GM, Chuang SE, Lee LM. 2013. Mefloquine induces cell death in prostate cancer cells and provides a potential novel treatment strategy in vivo. Oncol Lett **5**:1567–1571.

22 Liu Y, Chen S, Xue R, Zhao J, Di M. 2016. Mefloquine effectively targets gastric cancer cells through phosphatase-dependent inhibition of PI3K/Akt/mTOR signaling pathway. Biochem Biophys Res Commun **470**:350–355.

23 Rodrigues FA, Bomfim Ida S, Cavalcanti BC, Pessoa C, Goncalves RS, Wardell JL, Wardell SM, de Souza MV. 2014. Mefloquine-oxazolidine derivatives: a new class of anticancer agents. Chem Biol Drug Des **83**:126–131.

24 Krieger D, Vesenbeckh S, Schonfeld N, Bettermann G, Bauer TT, Russmann H, Mauch H. 2015. Mefloquine as a potential drug against multidrug-resistant tuberculosis. Eur Respir J **46**:1503–1505.

25 Bermudez LE, Meek L. 2014. Mefloquine and Its Enantiomers Are Active against Mycobacterium tuberculosis In Vitro and in Macrophages. Tuberc Res Treat **2014**:530815.

26 Goncalves RS, Kaiser CR, Lourenco MC, Bezerra FA, de Souza MV, Wardell JL, Wardell SM, Henriques M, Costa T. 2012. Mefloquine-oxazolidine derivatives, derived from mefloquine and arenecarbaldehydes: In vitro activity including against the multidrug-resistant tuberculosis strain T113. Bioorg Med Chem **20**:243–248.

27 Shin JW, Jung KH, Lee ST, Moon J, Lim JA, Byun JI, Park KI, Lee SK, Chu K. 2014. Mefloquine improved progressive multifocal leukoencephalopathy in a patient with immunoglobulin A nephropathy. J Clin Neurosci **21**:1661–1664.

28 Kurmann R, Weisstanner C, Kardas P, Hirsch HH, Wiest R, Lammle B, Furrer H, Du Pasquier R, Bassetti CL, Sturzenegger M, Krestel H. 2015. Progressive multifocal leukoencephalopathy in common variable immunodeficiency: mitigated course under mirtazapine and mefloquine. J Neurovirol **21**:694–701.

29 Gourineni VC, Juvet T, Kumar Y, Bordea D, Sena KN. 2014. Progressive multifocal leukoencephalopathy in a 62-year-old immunocompetent woman. Case Rep Neurol Med **2014**:549271.

30 Naito K, Ueno H, Sekine M, Kanemitsu M, Ohshita T, Nakamura T, Yamawaki T, Matsumoto M. 2012. Akinetic mutism caused by HIV-associated progressive multifocal leukoencephalopathy was successfully treated with mefloquine: a serial multimodal MRI Study. Intern Med **51**:205–209.

31 Barrows NJ, Campos RK, Powell ST, Prasanth KR, Schott-Lerner G, Soto-Acosta R, Galarza-Munoz G, McGrath EL, Urrabaz-Garza R, Gao J, Wu P, Menon R, Saade G, Fernandez-Salas I, Rossi SL, Vasilakis N, Routh A, Bradrick SS, Garcia-Blanco MA. 2016. A Screen of FDA-Approved Drugs for Inhibitors of Zika Virus Infection. Cell Host Microbe **20**:259–270.

32 Schlagenhauf P, Blumentals WA, Suter P, Regep L, Vital-Durand G, Schaerer MT, Boutros MS, Rhein HG, Adamcova M. 2012. Pregnancy and fetal outcomes after exposure to mefloquine in the pre- and periconception period and during pregnancy. Clin Infect Dis **54**:e124–131.

33 Karbwang J, White NJ. 1990. Clinical pharmacokinetics of mefloquine. Clin Pharmacokinet **19**:264–279.

34 Barbosa-Lima G, Moraes AM, Araujo AD, da Silva ET, de Freitas CS, Vieira YR, Marttorelli A, Neto JC, Bozza PT, de Souza MV, Souza TM. 2017. 2,8-bis(trifluoromethyl)quinoline analogs show improved anti-Zika virus activity, compared to mefloquine. Eur J Med Chem **127**:334–340.

35 Aman MJ, Kashanchi F. 2016. Zika Virus: A New Animal Model for an Arbovirus. PLoS Negl Trop Dis **10**:e0004702.

36 Dowall SD, Graham VA, Rayner E, Atkinson B, Hall G, Watson RJ, Bosworth A, Bonney LC, Kitchen S, Hewson R. 2016. A Susceptible Mouse Model for Zika Virus Infection. PLoS Negl Trop Dis **10**:e0004658.

37 Lazear HM, Govero J, Smith AM, Platt DJ, Fernandez E, Miner JJ, Diamond MS. 2016. A Mouse Model of Zika Virus Pathogenesis. Cell Host Microbe **19**:720–730.

38 Denisova OV, Kakkola L, Feng L, Stenman J, Nagaraj A, Lampe J, Yadav B, Aittokallio T, Kaukinen P, Ahola T, Kuivanen S, Vapalahti O, Kantele A, Tynell J, Julkunen I, Kallio-Kokko H, Paavilainen H, Hukkanen V, Elliott RM, De Brabander JK, Saelens X, Kainov DE.

2012. Obatoclax, saliphenylhalamide, and gemcitabine inhibit influenza a virus infection. J Biol Chem **287**:35324–35332.

39 Muller KH, Kainov DE, El Bakkouri K, Saelens X, De Brabander JK, Kittel C, Samm E, Muller CP. 2011. The proton translocation domain of cellular vacuolar ATPase provides a target for the treatment of influenza A virus infections. Br J Pharmacol **164**:344–357.

40 Muller KH, Spoden GA, Scheffer KD, Brunnhofer R, De Brabander JK, Maier ME, Florin L, Muller CP. 2014. Inhibition by cellular vacuolar ATPase impairs human papillomavirus uncoating and infection. Antimicrob Agents Chemother **58**:2905–2911.

41 Soderholm S, Anastasina M, Islam MM, Tynell J, Poranen MM, Bamford DH, Stenman J, Julkunen I, Sauliene I, De Brabander JK, Matikainen S, Nyman TA, Saelens X, Kainov D. 2016. Immuno-modulating properties of saliphenylhalamide, SNS-032, obatoclax, and gemcitabine. Antiviral Res **126**:69–80.

42 Konopleva M, Watt J, Contractor R, Tsao T, Harris D, Estrov Z, Bornmann W, Kantarjian H, Viallet J, Samudio I, Andreeff M. 2008. Mechanisms of antileukemic activity of the novel Bcl-2 homology domain-3 mimetic GX15-070 (obatoclax). Cancer Res **68**:3413–3420.

43 Miner JJ, Sene A, Richner JM, Smith AM, Santeford A, Ban N, Weger-Lucarelli J, Manzella F, Ruckert C, Govero J, Noguchi KK, Ebel GD, Diamond MS, Apte RS. 2016. Zika Virus Infection in Mice Causes Panuveitis with Shedding of Virus in Tears. Cell Rep **16**:3208–3218.

44 Kuivanen S, Bespalov MM, Nandania J, Ianevski A, Velagapudi V, De Brabander JK, Kainov DE, Vapalahti O. 2017. Obatoclax, saliphenylhalamide and gemcitabine inhibit Zika virus infection in vitro and differentially affect cellular signaling, transcription and metabolism. Antiviral Res **139**:117–128.

45 Driggers RW, Ho CY, Korhonen EM, Kuivanen S, Jaaskelainen AJ, Smura T, Rosenberg A, Hill DA, DeBiasi RL, Vezina G, Timofeev J, Rodriguez FJ, Levanov L, Razak J, Iyengar P, Hennenfent A, Kennedy R, Lanciotti R, **du** Plessis A, Vapalahti O. 2016. Zika Virus Infection with Prolonged Maternal Viremia and Fetal Brain Abnormalities. N Engl J Med **374**:2142–2151.

46 WHO model formulary 2008. World Health Organization, Geneva. http://BZ6FJ9FL8E.search.serialssolutions.com/?V=1.0&L=BZ6FJ9F L8E&S=JCs&C=TC_024266734&T=marc http://KG6EK7CQ2B. search.serialssolutions.com/?V=1.0&L=KG6EK7CQ2B&S=JCs&C=

TC_024266734&T=marc http://MN3KU4AR6Y.search.
serialssolutions.com/?v=1.0&L=MN3KU4AR6Y&S=JCs&C=TC_024
266734&T=marc http://RX8KL6YF4X.search.serialssolutions.com/?
V=1.0&L=RX8KL6YF4X&S=JCs&C=TC_024266734&T=marc
http://TE6UZ4HK6Z.search.serialssolutions.com/?v=1.0&L=TE6UZ
4HK6Z&S=JCs&C=TC_024266734&T=marc.

47 Andrews P, Thyssen J, Lorke D. 1982. The biology and toxicology of
    molluscicides, Bayluscide. Pharmacol Ther **19**:245–295.

48 Al-Hadiya BM. 2005. Niclosamide: comprehensive profile. Profiles
    Drug Subst Excip Relat Methodol **32**:67–96.

49 Wu CJ, Jan JT, Chen CM, Hsieh HP, Hwang DR, Liu HW, Liu CY,
    Huang HW, Chen SC, Hong CF, Lin RK, Chao YS, Hsu JT. 2004.
    Inhibition of severe acute respiratory syndrome coronavirus replication
    by niclosamide. Antimicrob Agents Chemother **48**:2693–2696.

50 Jurgeit A, McDowell R, Moese S, Meldrum E, Schwendener R,
    Greber UF. 2012. Niclosamide is a proton carrier and targets
    acidic endosomes with broad antiviral effects. PLoS Pathog
    8:e1002976.

51 Fang J, Sun L, Peng G, Xu J, Zhou R, Cao S, Chen H, Song Y. 2013.
    Identification of three antiviral inhibitors against Japanese
    encephalitis virus from library of pharmacologically active
    compounds 1280. PLoS One **8**:e78425.

52 Xu M, Lee EM, Wen Z, Cheng Y, Huang WK, Qian X, Tcw J,
    Kouznetsova J, Ogden SC, Hammack C, Jacob F, Nguyen HN, Itkin
    M, Hanna C, Shinn P, Allen C, Michael SG, Simeonov A, Huang W,
    Christian KM, Goate A, Brennand KJ, Huang R, Xia M, Ming GL,
    Zheng W, Song H, Tang H. 2016. Identification of small-molecule
    inhibitors of Zika virus infection and induced neural cell death via a
    drug repurposing screen. Nat Med **22**:1101–1107.

53 Khanim FL, Merrick BA, Giles HV, Jankute M, Jackson JB, Giles LJ,
    Birtwistle J, Bunce CM, Drayson MT. 2011. Redeployment-based
    drug screening identifies the anti-helminthic niclosamide as anti-
    myeloma therapy that also reduces free light chain production. Blood
    Cancer J **1**:e39.

54 Perachon S, Schwartz JC, Sokoloff P. 1999. Functional potencies of
    new antiparkinsonian drugs at recombinant human dopamine D1,
    D2 and D3 receptors. Eur J Pharmacol **366**:293–300.

55 Zhang Y, Scislowski PW, Prevelige R, Phaneuf S, Cincotta AH. 1999.
    Bromocriptine/SKF38393 treatment ameliorates dyslipidemia in ob/
    ob mice. Metabolism **48**:1033–1040.

56 Ginther OJ, Santos VG, Mir RA, Beg MA. 2012. Role of LH in the progesterone increase during the bromocriptine-induced prolactin decrease in heifers. Theriogenology **78**:1969–1976.

57 Kato F, Ishida Y, Oishi S, Fujii N, Watanabe S, Vasudevan SG, Tajima S, Takasaki T, Suzuki Y, Ichiyama K, Yamamoto N, Yoshii K, Takashima I, Kobayashi T, Miura T, Igarashi T, Hishiki T. 2016. Novel antiviral activity of bromocriptine against dengue virus replication. Antiviral Res **131**:141–147.

58 Chan JF, Chik KK, Yuan S, Yip CC, Zhu Z, Tee KM, Tsang JO, Chan CC, Poon VK, Lu G, Zhang AJ, Lai KK, Chan KH, Kao RY, Yuen KY. 2017. Novel antiviral activity and mechanism of bromocriptine as a Zika virus NS2B-NS3 protease inhibitor. Antiviral Res **141**:29–37.

59 Flogstad AK, Halse J, Grass P, Abisch E, Djoseland O, Kutz K, Bodd E, Jervell J. 1994. A comparison of octreotide, bromocriptine, or a combination of both drugs in acromegaly. J Clin Endocrinol Metab **79**:461–465.

60 del Pozo E, Schluter K, Nuesch E, Rosenthaler J, Kerp L. 1986. Pharmacokinetics of a long-acting bromocriptine preparation (Parlodel LA) and its effect on release of prolactin and growth hormone. Eur J Clin Pharmacol **29**:615–618.

61 Plockinger U, Quabbe HJ. 1991. Evaluation of a repeatable depot-bromocriptine preparation(Parlodel LAR) for the treatment of acromegaly. J Endocrinol Invest **14**:943–948.

62 Darwish AM, Hafez E, El-Gebali I, Hassan SB. 2005. Evaluation of a novel vaginal bromocriptine mesylate formulation: a pilot study. Fertil Steril **83**:1053–1055.

63 Darwish AM, El-Sayed AM, El-Harras SA, Khaled KA, Ismail MA. 2008. Clinical efficacy of novel unidirectional buccoadhesive vs. vaginoadhesive bromocriptine mesylate discs for treating pathologic hyperprolactinemia. Fertil Steril **90**:1864–1868.

64 Shiryaev SA, Ratnikov BI, Chekanov AV, Sikora S, Rozanov DV, Godzik A, Wang J, Smith JW, Huang Z, Lindberg I, Samuel MA, Diamond MS, Strongin AY. 2006. Cleavage targets and the D-arginine-based inhibitors of the West Nile virus NS3 processing proteinase. Biochem J **393**:503–511.

65 Chen X, Yang K, Wu C, Chen C, Hu C, Buzovetsky O, Wang Z, Ji X, Xiong Y, Yang H. 2016. Mechanisms of activation and inhibition of Zika virus NS2B-NS3 protease. Cell Res **26**:1260–1263.

66 Phoo WW, Li Y, Zhang Z, Lee MY, Loh YR, Tan YB, Ng EY, Lescar J, Kang C, Luo D. 2016. Structure of the NS2B-NS3 protease from Zika virus after self-cleavage. Nat Commun **7**:13410.

67 Lei J, Hansen G, Nitsche C, Klein CD, Zhang L, Hilgenfeld R. 2016. Crystal structure of Zika virus NS2B-NS3 protease in complex with a boronate inhibitor. Science **353**:503–505.

68 Roy A, Lim L, Srivastava S, Song J. 2016. Unique properties of Zika NS2B-NS3pro complexes as decoded by experiments and MD simulations. bioRxiv doi:10.1101/078113.

69 Lim L, Roy A, Song J. 2016. Identification of a Zika NS2B-NS3pro pocket susceptible to allosteric inhibition by small molecules including quercin rich in edible plants. bioRxiv doi:10.1101/078543.

70 Roy A, Lim L, Song J. 2016. Identification of quercetin from fruits to immediately fight Zika. bioRxiv doi:10.1101/074559.

71 Ekonomiuk D, Su XC, Ozawa K, Bodenreider C, Lim SP, Otting G, Huang D, Caflisch A. 2009. Flaviviral protease inhibitors identified by fragment-based library docking into a structure generated by molecular dynamics. J Med Chem **52**:4860–4868.

72 Sahoo M, Jena L, Daf S, Kumar S. 2016. Virtual Screening for Potential Inhibitors of NS3 Protein of Zika Virus. Genomics Inform **14**:104–111.

73 Bollati M, Milani M, Mastrangelo E, Ricagno S, Tedeschi G, Nonnis S, Decroly E, Selisko B, de Lamballerie X, Coutard B, Canard B, Bolognesi M. 2009. Recognition of RNA cap in the Wesselsbron virus NS5 methyltransferase domain: implications for RNA-capping mechanisms in Flavivirus. J Mol Biol **385**:140–152.

74 Egloff MP, Decroly E, Malet H, Selisko B, Benarroch D, Ferron F, Canard B. 2007. Structural and functional analysis of methylation and 5′-RNA sequence requirements of short capped RNAs by the methyltransferase domain of dengue virus NS5. J Mol Biol **372**:723–736.

75 Choi KH, Groarke JM, Young DC, Kuhn RJ, Smith JL, Pevear DC, Rossmann MG. 2004. The structure of the RNA-dependent RNA polymerase from bovine viral diarrhea virus establishes the role of GTP in de novo initiation. Proc Natl Acad Sci U S A **101**:4425–4430.

76 Geiss BJ, Thompson AA, Andrews AJ, Sons RL, Gari HH, Keenan SM, Peersen OB. 2009. Analysis of flavivirus NS5 methyltransferase cap binding. J Mol Biol **385**:1643–1654.

77 Grun JB, Brinton MA. 1986. Characterization of West Nile virus RNA-dependent RNA polymerase and cellular terminal adenylyl and uridylyl transferases in cell-free extracts. J Virol **60**:1113–1124.

78 Issur M, Geiss BJ, Bougie I, Picard-Jean F, Despins S, Mayette J, Hobdey SE, Bisaillon M. 2009. The flavivirus NS5 protein is a true

RNA guanylyltransferase that catalyzes a two-step reaction to form the RNA cap structure. RNA **15**:2340–2350.

79 Tan BH, Fu J, Sugrue RJ, Yap EH, Chan YC, Tan YH. 1996. Recombinant dengue type 1 virus NS5 protein expressed in Escherichia coli exhibits RNA-dependent RNA polymerase activity. Virology **216**:317–325.

80 Zou G, Chen YL, Dong H, Lim CC, Yap LJ, Yau YH, Shochat SG, Lescar J, Shi PY. 2011. Functional analysis of two cavities in flavivirus NS5 polymerase. J Biol Chem **286**:14362–14372.

81 Zonneveld R, Roosblad J, Staveren JW, Wilschut JC, Vreden SG, Codrington J. 2016. Three atypical lethal cases associated with acute Zika virus infection in Suriname. IDCases **5**:49–53.

82 Keating GM, Vaidya A. 2014. Sofosbuvir: first global approval. Drugs **74**:273–282.

83 Sofia MJ, Bao D, Chang W, Du J, Nagarathnam D, Rachakonda S, Reddy PG, Ross BS, Wang P, Zhang HR, Bansal S, Espiritu C, Keilman M, Lam AM, Steuer HM, Niu C, Otto MJ, Furman PA. 2010. Discovery of a beta-d-2′-deoxy-2′-alpha-fluoro-2′-beta-C-methyluridine nucleotide prodrug (PSI-7977) for the treatment of hepatitis C virus. J Med Chem **53**:7202–7218.

84 Murakami E, Tolstykh T, Bao H, Niu C, Steuer HM, Bao D, Chang W, Espiritu C, Bansal S, Lam AM, Otto MJ, Sofia MJ, Furman PA. 2010. Mechanism of activation of PSI-7851 and its diastereoisomer PSI-7977. J Biol Chem **285**:34337–34347.

85 Bhatia HK, Singh H, Grewal N, Natt NK. 2014. Sofosbuvir: A novel treatment option for chronic hepatitis C infection. J Pharmacol Pharmacother **5**:278–284.

86 Sacramento CQ, de Melo GR, de Freitas CS, Rocha N, Hoelz LV, Miranda M, Fintelman-Rodrigues N, Marttorelli A, Ferreira AC, Barbosa-Lima G, Abrantes JL, Vieira YR, Bastos MM, de Mello Volotao E, Nunes EP, Tschoeke DA, Leomil L, Loiola EC, Trindade P, Rehen SK, Bozza FA, Bozza PT, Boechat N, Thompson FL, de Filippis AM, Bruning K, Souza TM. 2017. The clinically approved antiviral drug sofosbuvir inhibits Zika virus replication. Sci Rep **7**:40920.

87 Bullard-Feibelman KM, Govero J, Zhu Z, Salazar V, Veselinovic M, Diamond MS, Geiss BJ. 2017. The FDA-approved drug sofosbuvir inhibits Zika virus infection. Antiviral Res **137**:134–140.

88 Onorati M, Li Z, Liu F, Sousa AM, Nakagawa N, Li M, Dell'Anno MT, Gulden FO, Pochareddy S, Tebbenkamp AT, Han W, Pletikos M, Gao T,

Zhu Y, Bichsel C, Varela L, Szigeti-Buck K, Lisgo S, Zhang Y, Testen A, Gao XB, Mlakar J, Popovic M, Flamand M, Strittmatter SM, Kaczmarek LK, Anton ES, Horvath TL, Lindenbach BD, Sestan N. 2016. Zika Virus Disrupts Phospho-TBK1 Localization and Mitosis in Human Neuroepithelial Stem Cells and Radial Glia. Cell Rep **16**:2576–2592.

89 Cada DJ, Cong J, Baker DE. 2014. Sofosbuvir. Hosp Pharm **49**:466–478.

90 Warren TK, Wells J, Panchal RG, Stuthman KS, Garza NL, Van Tongeren SA, Dong L, Retterer CJ, Eaton BP, Pegoraro G, Honnold S, Bantia S, Kotian P, Chen X, Taubenheim BR, Welch LS, Minning DM, Babu YS, Sheridan WP, Bavari S. 2014. Protection against filovirus diseases by a novel broad-spectrum nucleoside analogue BCX4430. Nature **508**:402–405.

91 Julander JG, Bantia S, Taubenheim BR, Minning DM, Kotian P, Morrey JD, Smee DF, Sheridan WP, Babu YS. 2014. BCX4430, a novel nucleoside analog, effectively treats yellow fever in a Hamster model. Antimicrob Agents Chemother **58**:6607–6614.

92 Julander JG, Siddharthan V, Evans J, Taylor R, Tolbert K, Apuli C, Stewart J, Collins P, Gebre M, Neilson S, Van Wettere A, Lee YM, Sheridan WP, Morrey JD, Babu YS. 2017. Efficacy of the broad-spectrum antiviral compound BCX4430 against Zika virus in cell culture and in a mouse model. Antiviral Res **137**:14–22.

93 Morrey JD, Smee DF, Sidwell RW, Tseng C. 2002. Identification of active antiviral compounds against a New York isolate of West Nile virus. Antiviral Res **55**:107–116.

94 Julander JG, Shafer K, Smee DF, Morrey JD, Furuta Y. 2009. Activity of T-705 in a hamster model of yellow fever virus infection in comparison with that of a chemically related compound, T-1106. Antimicrob Agents Chemother **53**:202–209.

95 Yin Z, Chen YL, Schul W, Wang QY, Gu F, Duraiswamy J, Kondreddi RR, Niyomrattanakit P, Lakshminarayana SB, Goh A, Xu HY, Liu W, Liu B, Lim JY, Ng CY, Qing M, Lim CC, Yip A, Wang G, Chan WL, Tan HP, Lin K, Zhang B, Zou G, Bernard KA, Garrett C, Beltz K, Dong M, Weaver M, He H, Pichota A, Dartois V, Keller TH, Shi PY. 2009. An adenosine nucleoside inhibitor of dengue virus. Proc Natl Acad Sci U S A **106**:20435–20439.

96 Lo MK, Shi PY, Chen YL, Flint M, Spiropoulou CF. 2016. In vitro antiviral activity of adenosine analog NITD008 against tick-borne flaviviruses. Antiviral Res **130**:46–49.

97 Qing J, Luo R, Wang Y, Nong J, Wu M, Shao Y, Tang R, Yu X, Yin Z, Sun Y. 2016. Resistance analysis and characterization of NITD008 as

an adenosine analog inhibitor against hepatitis C virus. Antiviral Res **126**:43–54.

98 Deng CL, Yeo H, Ye HQ, Liu SQ, Shang BD, Gong P, Alonso S, Shi PY, Zhang B. 2014. Inhibition of enterovirus 71 by adenosine analog NITD008. J Virol **88**:11915–11923.

99 Haddow AD, Schuh AJ, Yasuda CY, Kasper MR, Heang V, Huy R, Guzman H, Tesh RB, Weaver SC. 2012. Genetic characterization of Zika virus strains: geographic expansion of the Asian lineage. PLoS Negl Trop Dis **6**:e1477.

100 Deng YQ, Zhang NN, Li CF, Tian M, Hao JN, Xie XP, Shi PY, Qin CF. 2016. Adenosine Analog NITD008 Is a Potent Inhibitor of Zika Virus. Open Forum Infect Dis **3**:ofw175.

101 Olsen DB, Eldrup AB, Bartholomew L, Bhat B, Bosserman MR, Ceccacci A, Colwell LF, Fay JF, Flores OA, Getty KL, Grobler JA, LaFemina RL, Markel EJ, Migliaccio G, Prhavc M, Stahlhut MW, Tomassini JE, MacCoss M, Hazuda DJ, Carroll SS. 2004. A 7-deaza-adenosine analog is a potent and selective inhibitor of hepatitis C virus replication with excellent pharmacokinetic properties. Antimicrob Agents Chemother **48**:3944–3953.

102 Carroll SS, Ludmerer S, Handt L, Koeplinger K, Zhang NR, Graham D, Davies ME, MacCoss M, Hazuda D, Olsen DB. 2009. Robust antiviral efficacy upon administration of a nucleoside analog to hepatitis C virus-infected chimpanzees. Antimicrob Agents Chemother **53**:926–934.

103 Olsen DB, Davies ME, Handt L, Koeplinger K, Zhang NR, Ludmerer SW, Graham D, Liverton N, MacCoss M, Hazuda D, Carroll SS. 2011. Sustained viral response in a hepatitis C virus-infected chimpanzee via a combination of direct-acting antiviral agents. Antimicrob Agents Chemother **55**:937–939.

104 Arnold JJ, Sharma SD, Feng JY, Ray AS, Smidansky ED, Kireeva ML, Cho A, Perry J, Vela JE, Park Y, Xu Y, Tian Y, Babusis D, Barauskus O, Peterson BR, Gnatt A, Kashlev M, Zhong W, Cameron CE. 2012. Sensitivity of mitochondrial transcription and resistance of RNA polymerase II dependent nuclear transcription to antiviral ribonucleosides. PLoS Pathog **8**:e1003030.

105 Schul W, Liu W, Xu HY, Flamand M, Vasudevan SG. 2007. A dengue fever viremia model in mice shows reduction in viral replication and suppression of the inflammatory response after treatment with antiviral drugs. J Infect Dis **195**:665–674.

106 Eyer L, Valdes JJ, Gil VA, Nencka R, Hrebabecky H, Sala M, Salat J, Cerny J, Palus M, De Clercq E, Ruzek D. 2015. Nucleoside inhibitors

of tick-borne encephalitis virus. Antimicrob Agents Chemother **59**:5483–5493.

107 Wu R, Smidansky ED, Oh HS, Takhampunya R, Padmanabhan R, Cameron CE, Peterson BR. 2010. Synthesis of a 6-methyl-7-deaza analogue of adenosine that potently inhibits replication of polio and dengue viruses. J Med Chem **53**:7958–7966.

108 Goris N, De Palma A, Toussaint JF, Musch I, Neyts J, De Clercq K. 2007. 2′-C-methylcytidine as a potent and selective inhibitor of the replication of foot-and-mouth disease virus. Antiviral Res **73**:161–168.

109 Zmurko J, Marques RE, Schols D, Verbeken E, Kaptein SJ, Neyts J. 2016. The Viral Polymerase Inhibitor 7-Deaza-2′-C-Methyladenosine Is a Potent Inhibitor of In Vitro Zika Virus Replication and Delays Disease Progression in a Robust Mouse Infection Model. PLoS Negl Trop Dis **10**:e0004695.

110 Eyer L, Nencka R, Huvarova I, Palus M, Joao Alves M, Gould EA, De Clercq E, Ruzek D. 2016. Nucleoside Inhibitors of Zika Virus. J Infect Dis **214**:707–711.

111 Hercik K, Kozak J, Sala M, Dejmek M, Hrebabecky H, Zbornikova E, Smola M, Ruzek D, Nencka R, Boura E. 2017. Adenosine triphosphate analogs can efficiently inhibit the Zika virus RNA-dependent RNA polymerase. Antiviral Res **137**:131–133.

112 Kuhn RJ, Zhang W, Rossmann MG, Pletnev SV, Corver J, Lenches E, Jones CT, Mukhopadhyay S, Chipman PR, Strauss EG, Baker TS, Strauss JH. 2002. Structure of dengue virus: implications for flavivirus organization, maturation, and fusion. Cell **108**:717–725.

113 Zhang X, Ge P, Yu X, Brannan JM, Bi G, Zhang Q, Schein S, Zhou ZH. 2013. Cryo-EM structure of the mature dengue virus at 3.5-A resolution. Nat Struct Mol Biol **20**:105–110.

114 Sirohi D, Chen Z, Sun L, Klose T, Pierson TC, Rossmann MG, Kuhn RJ. 2016. The 3.8 A resolution cryo-EM structure of Zika virus. Science **352**:467–470.

115 Beltramello M, Williams KL, Simmons CP, Macagno A, Simonelli L, Quyen NT, Sukupolvi-Petty S, Navarro-Sanchez E, Young PR, de Silva AM, Rey FA, Varani L, Whitehead SS, Diamond MS, Harris E, Lanzavecchia A, Sallusto F. 2010. The human immune response to Dengue virus is dominated by highly cross-reactive antibodies endowed with neutralizing and enhancing activity. Cell Host Microbe **8**:271–283.

116 Oliphant T, Nybakken GE, Austin SK, Xu Q, Bramson J, Loeb M, Throsby M, Fremont DH, Pierson TC, Diamond MS. 2007.

Induction of epitope-specific neutralizing antibodies against West Nile virus. J Virol **81**:11828–11839.

117 Dejnirattisai W, Wongwiwat W, Supasa S, Zhang X, Dai X, Rouvinski A, Jumnainsong A, Edwards C, Quyen NT, Duangchinda T, Grimes JM, Tsai WY, Lai CY, Wang WK, Malasit P, Farrar J, Simmons CP, Zhou ZH, Rey FA, Mongkolsapaya J, Screaton GR. 2015. A new class of highly potent, broadly neutralizing antibodies isolated from viremic patients infected with dengue virus. Nat Immunol **16**:170–177.

118 Rouvinski A, Guardado-Calvo P, Barba-Spaeth G, Duquerroy S, Vaney MC, Kikuti CM, Navarro Sanchez ME, Dejnirattisai W, Wongwiwat W, Haouz A, Girard-Blanc C, Petres S, Shepard WE, Despres P, Arenzana-Seisdedos F, Dussart P, Mongkolsapaya J, Screaton GR, Rey FA. 2015. Recognition determinants of broadly neutralizing human antibodies against dengue viruses. Nature **520**:109–113.

119 Smith SA, de Alwis AR, Kose N, Harris E, Ibarra KD, Kahle KM, Pfaff JM, Xiang X, Doranz BJ, de Silva AM, Austin SK, Sukupolvi-Petty S, Diamond MS, Crowe JE, Jr. 2013. The potent and broadly neutralizing human dengue virus-specific monoclonal antibody 1C19 reveals a unique cross-reactive epitope on the bc loop of domain II of the envelope protein. MBio **4**:e00873–00813.

120 Deng YQ, Dai JX, Ji GH, Jiang T, Wang HJ, Yang HO, Tan WL, Liu R, Yu M, Ge BX, Zhu QY, Qin ED, Guo YJ, Qin CF. 2011. A broadly flavivirus cross-neutralizing monoclonal antibody that recognizes a novel epitope within the fusion loop of E protein. PLoS One **6**:e16059.

121 Lai CY, Williams KL, Wu YC, Knight S, Balmaseda A, Harris E, Wang WK. 2013. Analysis of cross-reactive antibodies recognizing the fusion loop of envelope protein and correlation with neutralizing antibody titers in Nicaraguan dengue cases. PLoS Negl Trop Dis **7**:e2451.

122 Sultana H, Foellmer HG, Neelakanta G, Oliphant T, Engle M, Ledizet M, Krishnan MN, Bonafe N, Anthony KG, Marasco WA, Kaplan P, Montgomery RR, Diamond MS, Koski RA, Fikrig E. 2009. Fusion loop peptide of the West Nile virus envelope protein is essential for pathogenesis and is recognized by a therapeutic cross-reactive human monoclonal antibody. J Immunol **183**:650–660.

123 Vogt MR, Dowd KA, Engle M, Tesh RB, Johnson S, Pierson TC, Diamond MS. 2011. Poorly neutralizing cross-reactive antibodies

against the fusion loop of West Nile virus envelope protein protect in vivo via Fcgamma receptor and complement-dependent effector mechanisms. J Virol **85**:11567–11580.

124 Dai L, Song J, Lu X, Deng YQ, Musyoki AM, Cheng H, Zhang Y, Yuan Y, Song H, Haywood J, Xiao H, Yan J, Shi Y, Qin CF, Qi J, Gao GF. 2016. Structures of the Zika Virus Envelope Protein and Its Complex with a Flavivirus Broadly Protective Antibody. Cell Host Microbe **19**:696–704.

125 Deng YQ, Zhao H, Li XF, Zhang NN, Liu ZY, Jiang T, Gu DY, Shi L, He JA, Wang HJ, Sun ZZ, Ye Q, Xie DY, Cao WC, Qin CF. 2016. Isolation, identification and genomic characterization of the Asian lineage Zika virus imported to China. Sci China Life Sci **59**:428–430.

126 Saborio P, Lanzas R, Arrieta G, Arguedas A. 1995. Paragonimus mexicanus pericarditis: report of two cases and review of the literature. J Trop Med Hyg **98**:316–318.

127 Cosme A, Ojeda E, Poch M, Bujanda L, Castiella A, Fernandez J. 2003. Sonographic findings of hepatic lesions in human fascioliasis. J Clin Ultrasound **31**:358–363.

128 Leonardi W, Zilbermintz L, Cheng LW, Zozaya J, Tran SH, Elliott JH, Polukhina K, Manasherob R, Li A, Chi X, Gharaibeh D, Kenny T, Zamani R, Soloveva V, Haddow AD, Nasar F, Bavari S, Bassik MC, Cohen SN, Levitin A, Martchenko M. 2016. Bithionol blocks pathogenicity of bacterial toxins, ricin, and Zika virus. Sci Rep **6**:34475.

129 Tang H, Hammack C, Ogden SC, Wen Z, Qian X, Li Y, Yao B, Shin J, Zhang F, Lee EM, Christian KM, Didier RA, Jin P, Song H, Ming GL. 2016. Zika Virus Infects Human Cortical Neural Progenitors and Attenuates Their Growth. Cell Stem Cell **18**:587–590.

130 Dang J, Tiwari SK, Lichinchi G, Qin Y, Patil VS, Eroshkin AM, Rana TM. 2016. Zika Virus Depletes Neural Progenitors in Human Cerebral Organoids through Activation of the Innate Immune Receptor TLR3. Cell Stem Cell **19**:258–265.

131 Wurzer WJ, Planz O, Ehrhardt C, Giner M, Silberzahn T, Pleschka S, Ludwig S. 2003. Caspase 3 activation is essential for efficient influenza virus propagation. EMBO J **22**:2717–2728.

132 Richard A, Tulasne D. 2012. Caspase cleavage of viral proteins, another way for viruses to make the best of apoptosis. Cell Death Dis **3**:e277.

133 Nwokolo C. 1974. Endemic paragonimiasis in Africa. Bull World Health Organ **50**:569–571.

134 Bono VH, Jr., Weissman SM, Frei E, 3rd. 1964. The Effect of 6-Azauridine Administration on De Novo Pyrimidine Production in Chronic Myelogenous Leukemia. J Clin Invest **43**:1486–1494.

135 Harmon MW, Janis B. 1976. Effects of cytosine arabinoside, adenine arabinoside, and 6–azauridine on rabies virus in vitro and in vivo. J Infect Dis **133**:7–13.

136 Adcock RS, Chu YK, Golden JE, Chung DH. 2017. Evaluation of anti-Zika virus activities of broad-spectrum antivirals and NIH clinical collection compounds using a cell-based, high-throughput screen assay. Antiviral Res **138**:47–56.

137 Chung DH, Golden JE, Adcock RS, Schroeder CE, Chu YK, Sotsky JB, Cramer DE, Chilton PM, Song C, Anantpadma M, Davey RA, Prodhan AI, Yin X, Zhang X. 2016. Discovery of a Broad-Spectrum Antiviral Compound That Inhibits Pyrimidine Biosynthesis and Establishes a Type 1 Interferon-Independent Antiviral State. Antimicrob Agents Chemother **60**:4552–4562.

138 Liu S, Neidhardt EA, Grossman TH, Ocain T, Clardy J. 2000. Structures of human dihydroorotate dehydrogenase in complex with antiproliferative agents. Structure **8**:25–33.

139 Lucas-Hourani M, Dauzonne D, Jorda P, Cousin G, Lupan A, Helynck O, Caignard G, Janvier G, Andre-Leroux G, Khiar S, Escriou N, Despres P, Jacob Y, Munier-Lehmann H, Tangy F, Vidalain PO. 2013. Inhibition of pyrimidine biosynthesis pathway suppresses viral growth through innate immunity. PLoS Pathog **9**:e1003678.

140 Bonavia A, Franti M, Pusateri Keaney E, Kuhen K, Seepersaud M, Radetich B, Shao J, Honda A, Dewhurst J, Balabanis K, Monroe J, Wolff K, Osborne C, Lanieri L, Hoffmaster K, Amin J, Markovits J, Broome M, Skuba E, Cornella-Taracido I, Joberty G, Bouwmeester T, Hamann L, Tallarico JA, Tommasi R, Compton T, Bushell SM. 2011. Identification of broad-spectrum antiviral compounds and assessment of the druggability of their target for efficacy against respiratory syncytial virus (RSV). Proc Natl Acad Sci U S A **108**:6739–6744.

141 Hoffmann HH, Kunz A, Simon VA, Palese P, Shaw ML. 2011. Broad-spectrum antiviral that interferes with de novo pyrimidine biosynthesis. Proc Natl Acad Sci U S A **108**:5777–5782.

142 Smee DF, Hurst BL, Day CW. 2012. D282, a non-nucleoside inhibitor of influenza virus infection that interferes with de novo pyrimidine biosynthesis. Antivir Chem Chemother **22**:263–272.

143 Wang QY, Bushell S, Qing M, Xu HY, Bonavia A, Nunes S, Zhou J, Poh MK, Florez de Sessions P, Niyomrattanakit P, Dong H, Hoffmaster K, Goh A, Nilar S, Schul W, Jones S, Kramer L, Compton T, Shi PY. 2011. Inhibition of dengue virus through suppression of host pyrimidine biosynthesis. J Virol **85**:6548–6556.

144 Schang LM, St Vincent MR, Lacasse JJ. 2006. Five years of progress on cyclin-dependent kinases and other cellular proteins as potential targets for antiviral drugs. Antivir Chem Chemother **17**:293–320.

145 Badia R, Angulo G, Riveira-Munoz E, Pujantell M, Puig T, Ramirez C, Torres-Torronteras J, Marti R, Pauls E, Clotet B, Ballana E, Este JA. 2016. Inhibition of herpes simplex virus type 1 by the CDK6 inhibitor PD-0332991 (palbociclib) through the control of SAMHD1. J Antimicrob Chemother **71**:387–394.

146 Nemeth G, Varga Z, Greff Z, Bencze G, Sipos A, Szantai-Kis C, Baska F, Gyuris A, Kelemenics K, Szathmary Z, Minarovits J, Keri G, Orfi L. 2011. Novel, selective CDK9 inhibitors for the treatment of HIV infection. Curr Med Chem **18**:342–358.

147 Okamoto M, Hidaka A, Toyama M, Hosoya T, Yamamoto M, Hagiwara M, Baba M. 2015. Selective inhibition of HIV-1 replication by the CDK9 inhibitor FIT-039. Antiviral Res **123**:1–4.

148 Van Duyne R, Guendel I, Jaworski E, Sampey G, Klase Z, Chen H, Zeng C, Kovalskyy D, El Kouni MH, Lepene B, Patanarut A, Nekhai S, Price DH, Kashanchi F. 2013. Effect of mimetic CDK9 inhibitors on HIV-1-activated transcription. J Mol Biol **425**:812–829.

149 Yamamoto M, Onogi H, Kii I, Yoshida S, Iida K, Sakai H, Abe M, Tsubota T, Ito N, Hosoya T, Hagiwara M. 2014. CDK9 inhibitor FIT-039 prevents replication of multiple DNA viruses. J Clin Invest **124**:3479–3488.

150 Zhang J, Li G, Ye X. 2010. Cyclin T1/CDK9 interacts with influenza A virus polymerase and facilitates its association with cellular RNA polymerase II. J Virol **84**:12619–12627.

151 Haddad JJ. 2013. Current opinion on 3-[2-[(2-tert-butyl-phenylaminooxalyl)-amino]-propionylamino]- 4-oxo-5-(2,3,5, 6-tetrafluoro-phenoxy)-pentanoic acid, an investigational drug targeting caspases and caspase-like proteases: the clinical trials in sight and recent anti-inflammatory advances. Recent Pat Inflamm Allergy Drug Discov 7:229–258.

152 Hoglen NC, Chen LS, Fisher CD, Hirakawa BP, Groessl T, Contreras PC. 2004. Characterization of IDN-6556 (3-[2-(2-tert-butyl-phenylaminooxalyl)-amino]-propionylamino]-4-oxo-5-(2,3,5,6-te

trafluoro-phenoxy)-pentanoic acid): a liver-targeted caspase inhibitor. J Pharmacol Exp Ther **309**:634–640.

153 Barreyro FJ, Holod S, Finocchietto PV, Camino AM, Aquino JB, Avagnina A, Carreras MC, Poderoso JJ, Gores GJ. 2015. The pancaspase inhibitor Emricasan (IDN-6556) decreases liver injury and fibrosis in a murine model of non-alcoholic steatohepatitis. Liver Int **35**:953–966.

154 Shiffman ML, Pockros P, McHutchison JG, Schiff ER, Morris M, Burgess G. 2010. Clinical trial: the efficacy and safety of oral PF-03491390, a pancaspase inhibitor—a randomized placebo-controlled study in patients with chronic hepatitis C. Aliment Pharmacol Ther **31**:969–978.

155 Panahi Y, Poursaleh Z, Goldust M. 2015. The efficacy of topical and oral ivermectin in the treatment of human scabies. Ann Parasitol **61**:11–16.

156 Lundberg L, Pinkham C, Baer A, Amaya M, Narayanan A, Wagstaff KM, Jans DA, Kehn-Hall K. 2013. Nuclear import and export inhibitors alter capsid protein distribution in mammalian cells and reduce Venezuelan Equine Encephalitis Virus replication. Antiviral Res **100**:662–672.

157 Varghese FS, Kaukinen P, Glasker S, Bespalov M, Hanski L, Wennerberg K, Kummerer BM, Ahola T. 2016. Discovery of berberine, abamectin and ivermectin as antivirals against chikungunya and other alphaviruses. Antiviral Res **126**:117–124.

158 Mastrangelo E, Pezzullo M, De Burghgraeve T, Kaptein S, Pastorino B, Dallmeier K, de Lamballerie X, Neyts J, Hanson AM, Frick DN, Bolognesi M, Milani M. 2012. Ivermectin is a potent inhibitor of flavivirus replication specifically targeting NS3 helicase activity: new prospects for an old drug. J Antimicrob Chemother **67**:1884–1894.

159 Eugui EM, Mirkovich A, Allison AC. 1991. Lymphocyte-selective antiproliferative and immunosuppressive activity of mycophenolic acid and its morpholinoethyl ester (RS-61443) in rodents. Transplant Proc **23**:15–18.

160 Eugui EM, Almquist SJ, Muller CD, Allison AC. 1991. Lymphocyte-selective cytostatic and immunosuppressive effects of mycophenolic acid in vitro: role of deoxyguanosine nucleotide depletion. Scand J Immunol **33**:161–173.

161 Diamond MS, Zachariah M, Harris E. 2002. Mycophenolic acid inhibits dengue virus infection by preventing replication of viral RNA. Virology **304**:211–221.

162 Kang S, Shields AR, Jupatanakul N, Dimopoulos G. 2014. Suppressing dengue-2 infection by chemical inhibition of Aedes aegypti host factors. PLoS Negl Trop Dis **8**:e3084.

163 Ng CY, Gu F, Phong WY, Chen YL, Lim SP, Davidson A, Vasudevan SG. 2007. Construction and characterization of a stable subgenomic dengue virus type 2 replicon system for antiviral compound and siRNA testing. Antiviral Res **76**:222–231.

164 Takhampunya R, Ubol S, Houng HS, Cameron CE, Padmanabhan R. 2006. Inhibition of dengue virus replication by mycophenolic acid and ribavirin. J Gen Virol **87**:1947–1952.

165 Streit JM, Jones RN, Sader HS. 2004. Daptomycin activity and spectrum: a worldwide sample of 6737 clinical Gram-positive organisms. J Antimicrob Chemother **53**:669–674.

166 Baltz RH. 2009. Daptomycin: mechanisms of action and resistance, and biosynthetic engineering. Curr Opin Chem Biol **13**:144–151.

167 Zaitseva E, Yang ST, Melikov K, Pourmal S, Chernomordik LV. 2010. Dengue virus ensures its fusion in late endosomes using compartment-specific lipids. PLoS Pathog **6**:e1001131.

168 Meyer JH, Wilson AA, Sagrati S, Hussey D, Carella A, Potter WZ, Ginovart N, Spencer EP, Cheok A, Houle S. 2004. Serotonin transporter occupancy of five selective serotonin reuptake inhibitors at different doses: an [11C]DASB positron emission tomography study. Am J Psychiatry **161**:826–835.

169 Matsuda S, Koyasu S. 2000. Mechanisms of action of cyclosporine. Immunopharmacology **47**:119–125.

170 Qing M, Yang F, Zhang B, Zou G, Robida JM, Yuan Z, Tang H, Shi PY. 2009. Cyclosporine inhibits flavivirus replication through blocking the interaction between host cyclophilins and viral NS5 protein. Antimicrob Agents Chemother **53**:3226–3235.

10

Long-Term Care and Perspectives

10.1 Prenatal Care and Diagnosis of Abnormal Fetus Development

From 2015 to 2017, there were 2,197 and 4,504 pregnant women that tested positive for Zika virus (ZIKV) in the United States, and US territories and freely associated states, respectively, as reported by Center for Disease Control and Prevention (CDC) in September 2017 (1). At least 98 pregnancies were affected by ZIKV in United States and the District of Columbia, and 138 in US territories and freely associated states.

It is currently recommended that pregnant women be tested for ZIKV infection in endemic areas, initially by reverse transcription polymerase-chain reaction (RT-PCR), and if positive, then by plaque-reduction neutralization testing (PRNT). Infection by ZIKV, which usually lasts 3 to 12 days, could occur at any time during pregnancy. Additionally, ZIKV IgM can be detected in the serum at as early as 4 days post-infection (d.p.i.) and up to 12 weeks after the development of symptoms (2, 3). Other samples, such as urine, can be used after the acute phase of infection. Pregnant women should also be screened for other viruses, such as dengue virus (DENV), chikungunya virus (CHIKV), and yellow fever virus (YFV), since they are transmitted by the same vector *Aedes ssp.*

Following a positive diagnosis for ZIKV, the next most important evaluation is by image diagnosis such as ultrasound and/or magnetic resonance imaging (MRI) to monitor the development of the fetus. Ultrasound has two advantages: it is an inexpensive method and can

Zika Virus and Diseases: From Molecular Biology to Epidemiology, First Edition.
Suzane Ramos da Silva, Fan Cheng and Shou-Jiang Gao.
© 2018 John Wiley & Sons, Inc. Published 2018 by John Wiley & Sons, Inc.

detect abnormal growth of the fetus at early stages. MRI is recommended after mid-trimester and more advanced pregnancy. In some rural areas, pregnant women have limited access to MRI testing because the hospitals or health centers that could provide such services are located far away from the basic health units. Another concerning problem, which has emerged since the report of the association of ZIKV infection with microcephaly, is that some women choose not to know if the fetus is having a development problem. Therefore, it is important that the mother has psychological assistance during the entire pregnancy if an abnormality is detected.

10.2 Long-Term Care for Patients Affected by ZIKV

The care for children impacted by ZIKV has been a challenge, especially in less-developed countries such as Brazil. The families receive public health care, but specialized hospitals that are able to treat the children are usually concentrated in the big cities far away from the families' residences. The challenge is to transport affected children to the health care center, and sometimes to multiple centers, to obtain the proper treatments or therapies.

Although the children present different levels of brain development, most of them might require care for their whole lives. The detection of microcephaly after birth had taken most of the families by surprise without proper preparation for the situation in 2015. The ZIKV epidemic was a surprise, and the most afflicted country, Brazil, was also unprepared, and was not able to treat sudden emergency of many children in a short period. Problems as lack of infrastructure for diagnosis, treatment, and psychological and financial support of the families are still present nowadays.

10.3 Assistance to Families with Children Affected by ZIKV

Families with children affected by microcephaly or other neurological problems can find assistance in non-profit organizations such as March of Dimes (marchofdimes.org), Health Children (healthchildren.org),

Center for Parent Information and Resources (parentcenterhub. org), Parent to Parent USA (p2pusa.org), Partnership for Parents (partnershipforparents.net), and Kids Health (kidshealth.org), to suggest some of them.

10.4 Perspectives

The methods of diagnosis of ZIKV infection have been significantly improved in the last two years (4). Early and precise diagnosis is essential for implementing appropriate measures of care. There are two main paths to effectively control ZIKV infection. An effective vaccine and the control of the *Aedes spp* mosquitoes.

The development of an effective vaccine is the ultimate goal for the control of ZIKV infection. In the last 2 years, most of the studies have described how the virus affects the cells, which cellular pathways are affected, and what existing drugs are effective for inhibiting ZIKV infection. These findings are essential for the development of an effective vaccine. The development of an infectious cDNA clone of ZIKV should facilitate further understanding of the biology of ZIKV infection and development of new drugs (5–7). Delineation of the functions of the viral genes and differences of the virus strains should help to identify the important coding regions of the viral genome that are essential for infection and pathogenesis (6, 8), and decrease transmission (9–11). Another important point is the co-infection of ZIKV and DENV in patients, which might alter the pathologies of ZIKV-associated diseases, increase pathogenesis of ZIKV infection, as well as influence the development of vaccines (12–19).

Studies had showed that some potential vaccines provide protection to mice and/or nonhuman primate (20–22), including one based on ZIKV NS1 protein, which was able to protect 100.0% of the immuno-competent mice after one low dose of vaccine (23). Recently, a vaccine based on ZIKV prME protein was able to prevent damage to the testes in immune-deficient mice, reducing the persistence of ZIKV in the testes (24). Currently, a vaccine is being tested in Central and South America as part of the clinical trial VRC 705. This vaccine, in phase 2/2b, was developed by National Institute of Allergy and Infectious Diseases (NIAID). Its safety, ability to stimulate the immune system, and dosage are being evaluated.

The second path is to focus on the control of mosquitoes. The use of insecticides might lead to the development of resistant population (25), hence demanding the development of new alternative approaches. Mosquitoes genetically modified (GM), produced by a British company Oxitec, have been used with success in Brazil, and more recently in Florida in the United States. The mating of the released male mosquitoes with the females would produce an inviable offspring, leading to a reduction of the population of *Aedes spp* by up to 90%. Interesting results showed that ZIKV isolated from America was better transmitted by *Aedes aegypti* than strains isolated in Asia (26), indicating that multiple factors should be taken in consideration to understand the differences in the pathogenesis and fitness between the different ZIKV strains.

An interesting development is the use of ZIKV for the treatment of glioblastoma, an aggressive brain cancer (27). Most drugs cannot cross the brain barrier, which has posed a challenge for glioblastoma treatment. The fact that ZIKV can cross the brain barrier indicates that it might be possible to generate a mutant for targeting cancer cells without causing any neurological diseases.

Additionally, almost daily, new studies generate new information about ZIKV. The search for new therapies against ZIKV is dynamic, and our lab has identified that PKI 14-22, a protein kinase A (PKA) inhibitor, as a potent inhibitor of ZIKV replication (28). Recently, reports have indicated that a natural plant from Australia might kill cells infected by ZIKV, and vaginal lactobacillus might decrease virus replication and sexual transmission. All recent knowledge on the outbreaks has further impacted the development of a vaccine and provided clues for developing effective therapeutic and preventive approaches against ZIKV infection.

References

1 CDC. Centers for Disease Control and Prevention. Pregnant Women with Any Laboratory Evidence of Possible Zika Virus Infection, 2015-2017. September 29, 2017. https://wwwcdcgov/zika/reporting/pregwomen-uscaseshtml.

2 Citil Dogan A, Wayne S, Bauer S, Ogunyemi D, Kulkharni SK, Maulik D, Carpenter CF, Bahado-Singh RO. 2017. The Zika virus and pregnancy: evidence, management, and prevention. J Matern Fetal Neonatal Med **30**:386–396.

3 Vouga M, Musso D, Van Mieghem T, Baud D. 2016. CDC guidelines for pregnant women during the Zika virus outbreak. Lancet **387**:843–844.

4 Wong SJ, Furuya A, Zou J, Xie X, Dupuis AP, 2nd, Kramer LD, Shi PY. 2017. A Multiplex Microsphere Immunoassay for Zika Virus Diagnosis. EBioMedicine **16**:136–140.

5 Shan C, Xie X, Muruato AE, Rossi SL, Roundy CM, Azar SR, Yang Y, Tesh RB, Bourne N, Barrett AD, Vasilakis N, Weaver SC, Shi PY. 2016. An Infectious cDNA Clone of Zika Virus to Study Viral Virulence, Mosquito Transmission, and Antiviral Inhibitors. Cell Host Microbe **19**:891–900.

6 Sapparapu G, Fernandez E, Kose N, Bin C, Fox JM, Bombardi RG, Zhao H, Nelson CA, Bryan AL, Barnes T, Davidson E, Mysorekar IU, Fremont DH, Doranz BJ, Diamond MS, Crowe JE. 2016. Neutralizing human antibodies prevent Zika virus replication and fetal disease in mice. Nature **540**:443–447.

7 Yang Y, Shan C, Zou J, Muruato AE, Bruno DN, de Almeida Medeiros Daniele B, Vasconcelos PF, Rossi SL, Weaver SC, Xie X, Shi PY. 2017. A cDNA Clone-Launched Platform for High-Yield Production of Inactivated Zika Vaccine. EBioMedicine doi:10.1016/j.ebiom.2017.02.003.

8 Bordon Y. 2016. Infection: Zika virus: end of transmission? Nat Rev Immunol **16**:718–719.

9 Aliota MT, Peinado SA, Velez ID, Osorio JE. 2016. The wMel strain of Wolbachia Reduces Transmission of Zika virus by Aedes aegypti. Sci Rep **6**:28792.

10 Melo CF, de Oliveira DN, Lima EO, Guerreiro TM, Esteves CZ, Beck RM, Padilla MA, Milanez GP, Arns CW, Proenca-Modena JL, Souza-Neto JA, Catharino RR. 2016. A Lipidomics Approach in the Characterization of Zika-Infected Mosquito Cells: Potential Targets for Breaking the Transmission Cycle. PLoS One **11**:e0164377.

11 Alam A, Ali S, Ahamad S, Malik MZ, Ishrat R. 2016. From ZikV genome to vaccine: in silico approach for the epitope-based peptide vaccine against Zika virus envelope glycoprotein. Immunology **149**:386–399.

12 Tang B, Xiao Y, Wu J. 2016. Implication of vaccination against dengue for Zika outbreak. Sci Rep **6**:35623.

13 Wen J, Tang WW, Sheets N, Ellison J, Sette A, Kim K, Shresta S. 2017. Identification of Zika virus epitopes reveals immunodominant and protective roles for dengue virus cross-reactive CD8+ T cells. Nat Microbiol **2**:17036.

14 Miner JJ, Diamond MS. 2017. Dengue Antibodies, then Zika: A Fatal Sequence in Mice. Immunity **46**:771–773.

15 Halstead SB. 2017. Biologic Evidence Required for Zika Disease Enhancement by Dengue Antibodies. Emerg Infect Dis **23**:569–573.

16 Castanha PMS, Nascimento EJM, Braga C, Cordeiro MT, de Carvalho OV, de Mendonca LR, Azevedo EAN, Franca RFO, Dhalia R, Marques ETA. 2017. Dengue Virus-Specific Antibodies Enhance Brazilian Zika Virus Infection. J Infect Dis **215**:781–785.

17 Priyamvada L, Quicke KM, Hudson WH, Onlamoon N, Sewatanon J, Edupuganti S, Pattanapanyasat K, Chokephaibulkit K, Mulligan MJ, Wilson PC, Ahmed R, Suthar MS, Wrammert J. 2016. Human antibody responses after dengue virus infection are highly cross-reactive to Zika virus. Proc Natl Acad Sci U S A **113**:7852–7857.

18 Paul LM, Carlin ER, Jenkins MM, Tan AL, Barcellona CM, Nicholson CO, Michael SF, Isern S. 2016. Dengue virus antibodies enhance Zika virus infection. Clin Transl Immunology **5**:e117.

19 Bardina SV, Bunduc P, Tripathi S, Duehr J, Frere JJ, Brown JA, Nachbagauer R, Foster GA, Krysztof D, Tortorella D, Stramer SL, Garcia-Sastre A, Krammer F, Lim JK. 2017. Enhancement of Zika virus pathogenesis by preexisting antiflavivirus immunity. Science **356**:175–180.

20 Abbink P, Larocca RA, De La Barrera RA, Bricault CA, Moseley ET, Boyd M, Kirilova M, Li Z, Ng'ang'a D, Nanayakkara O, Nityanandam R, Mercado NB, Borducchi EN, Agarwal A, Brinkman AL, Cabral C, Chandrashekar A, Giglio PB, Jetton D, Jimenez J, Lee BC, Mojta S, Molloy K, Shetty M, Neubauer GH, Stephenson KE, Peron JP, Zanotto PM, Misamore J, Finneyfrock B, Lewis MG, Alter G, Modjarrad K, Jarman RG, Eckels KH, Michael NL, Thomas SJ, Barouch DH. 2016. Protective efficacy of multiple vaccine platforms against Zika virus challenge in rhesus monkeys. Science **353**:1129–1132.

21 Pardi N, Hogan MJ, Pelc RS, Muramatsu H, Andersen H, DeMaso CR, Dowd KA, Sutherland LL, Scearce RM, Parks R, Wagner W, Granados A, Greenhouse J, Walker M, Willis E, Yu JS, McGee CE, Sempowski GD, Mui BL, Tam YK, Huang YJ, Vanlandingham D, Holmes VM, Balachandran H, Sahu S, Lifton M, Higgs S, Hensley SE, Madden TD, Hope MJ, Kariko K, Santra S, Graham BS, Lewis MG, Pierson TC, Haynes BF, Weissman D. 2017. Zika virus protection by a single low-dose nucleoside-modified mRNA vaccination. Nature **543**:248–251.

22 Dudley DM, Aliota MT, Mohr EL, Weiler AM, Lehrer-Brey G, Weisgrau KL, Mohns MS, Breitbach ME, Rasheed MN, Newman CM, Gellerup DD, Moncla LH, Post J, Schultz-Darken N, Schotzko ML, Hayes JM, Eudailey JA, Moody MA, Permar SR, O'Connor SL, Rakasz EG, Simmons HA, Capuano S, Golos TG, Osorio JE, Friedrich TC, O'Connor DH. 2016. A rhesus macaque model of Asian-lineage Zika virus infection. Nat Commun **7**:12204.

23 GeoVax Labs I, (Atlanta, GA). 2017. DEVELOPMENT OF A NOVEL VACCINE FOR ZIKA, ASM Communications - ASM Microbe 2017 meeting

24 Griffin BD, Muthumani K, Warner BM, Majer A, Hagan M, Audet J, Stein DR, Ranadheera C, Racine T, De La Vega MA, Piret J, Kucas S, Tran KN, Frost KL, De Graff C, Soule G, Scharikow L, Scott J, McTavish G, Smid V, Park YK, Maslow JN, Sardesai NY, Kim JJ, Yao XJ, Bello A, Lindsay R, Boivin G, Booth SA, Kobasa D, Embury-Hyatt C, Safronetz D, Weiner DB, Kobinger GP. 2017. DNA vaccination protects mice against Zika virus-induced damage to the testes. Nat Commun **8**:15743.

25 Hunter P. 2016. Challenges and options for disease vector control: The outbreak of Zika virus in South America and increasing insecticide resistance among mosquitoes have rekindled efforts for controlling disease vectors. EMBO Rep **17**:1370–1373.

26 Pompon J, Morales-Vargas R, Manuel M, Huat Tan C, Vial T, Hao Tan J, Sessions OM, Vasconcelos PdC, Ng LC, Missé D. 2017. A Zika virus from America is more efficiently transmitted than an Asian virus by Aedes aegypti mosquitoes from Asia. Scientific Reports **7**:1215.

27 Zhu Z, Gorman MJ, McKenzie LD, Chai JN, Hubert CG, Prager BC, Fernandez E, Richner JM, Zhang R, Shan C, Tycksen E, Wang X, Shi PY, Diamond MS, Rich JN, Chheda MG. 2017. Zika virus has oncolytic activity against glioblastoma stem cells. J Exp Med **214**:2843–2857.

28 Cheng F, Ramos da Silva, Huang IC, Jung JU, Gao SJ. 2017. Suppression of Zika virus infection and replication in endothelial cells and astrocytes by PKA inhibitor PKI 14-22. J Virol Dec 6. pii: JVI.02019-17.

Index

Zika Virus and Diseases: From Molecular Biology to Epidemiology, First Edition.
Suzane Ramos da Silva, Fan Cheng and Shou-Jiang Gao.
© 2018 John Wiley & Sons, Inc. Published 2018 by John Wiley & Sons, Inc.